Geohistory of the Izu Peninsula

MASATO KOYAMA

THE SHIZUOKA SHIMBUN

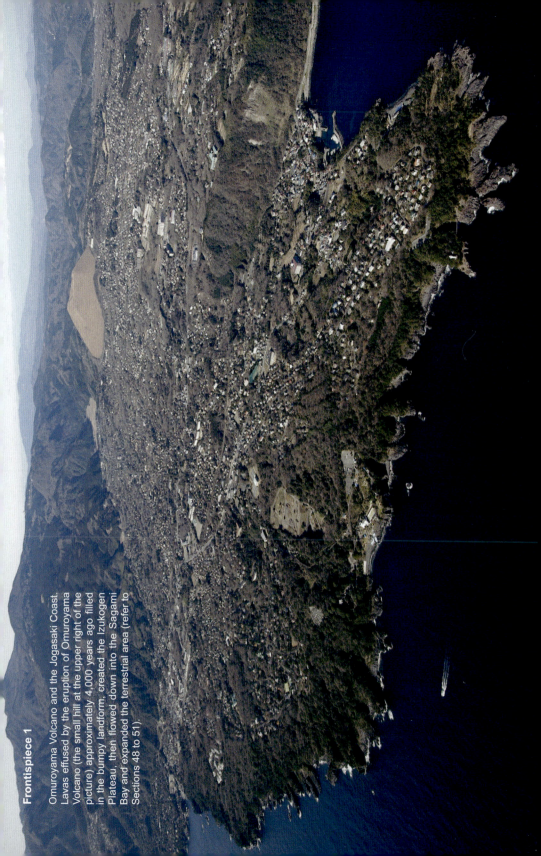

Frontispiece 1

Omuroyama Volcano and the Jogasaki Coast. Lavas effused by the eruption of Omuroyama Volcano (the small hill at the upper right of the picture) approximately 4,000 years ago filled in the bumpy landform, created the Izukogen Plateau, then flowed down into the Sagami Bay and expanded the terrestrial area (refer to Sections 48 to 51).

Frontispiece 2

A bird's-eye view of Omuroyama Volcano seen from its northwestern side. The picture shows a scoria cone that was created as a result of splashes of magma (scoria) having spouted in the air like a fountain and having piled up around the crater (refer to Section 48).

A bird's-eye view of Lake Ippekiko seen from its northwestern side. The picture shows a crater (maar) created as a result of phreatomagmatic explosion approximately 100,000 years ago (refer to Section 40).

Frontispiece 3

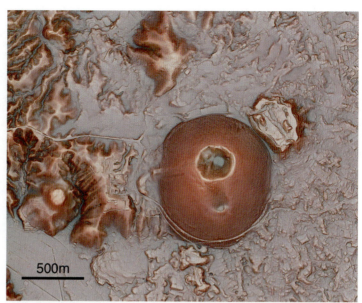

Omuroyama Volcano seen by Red Relief Image Map (RRIM). The picture presents a clear image of the mountain's beautiful circular shape and crater. The tableland northeast (at the upper right) of Omuroyama Volcano is Mt. Iwamuroyama (refer to Sections 48-49), which is one of the lava flow outlets. The image was provided by the Numazu River and National Highway Office of the Ministry of Land, Infrastructure, Transport and Tourism.

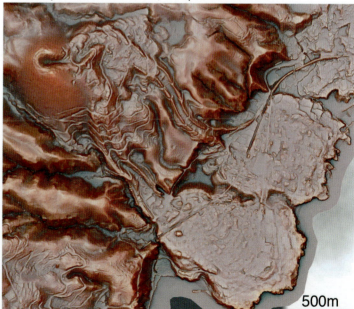

Red Relief Image Map (RRIM) of Ioyama Volcano and Ukiyama lava plateaus. The image map clearly shows that approximately 2,700 years ago, two streams of lava flowed down from Ioyama Volcano (at the upper left of the map) to the Sagami Bay and created two lava plateaus on the coast (refer to Section 55). The image was provided by the Numazu River and National Highway Office of the Ministry of Land, Infrastructure, Transport and Tourism.

Frontispiece 4

The columnar joints at Hashidate on the Jogasaki Coast, created by the lava flows effused from Omuroyama Volcano. The picture clearly shows the beautiful regular arrangement of many-sided prism-shaped fissures formed by the cooling and contraction of the lavas (refer to Section 50).

The pothole seen on the Kannonhama Beach along the Jogasaki Coast. A block of lava, washed and polished by waves, became a ball-shaped object while digging a hole in the rock around it. It is currently designated as a natural monument by the Ito municipal government (refer to Section 51).

Frontispiece 5

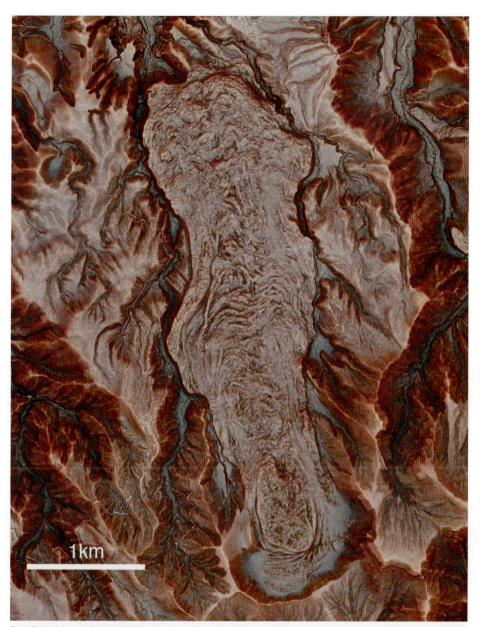

Red Relief Image Map (RRIM) of a lava flow effused from Kawagodaira Volcano. At the end of a huge eruption that occurred around the summit of Mt. Amagisan approximately 3,200 years ago, a high-viscosity lava flowed down northward from the crater (at the very bottom of the map) (refer to Section 52). There are many "wrinkles" created in the process of the northward drift of the lava flow. The image was provided by the Numazu River and National Highway Office of the Ministry of Land, Infrastructure, Transport and Tourism.

Frontispiece 6

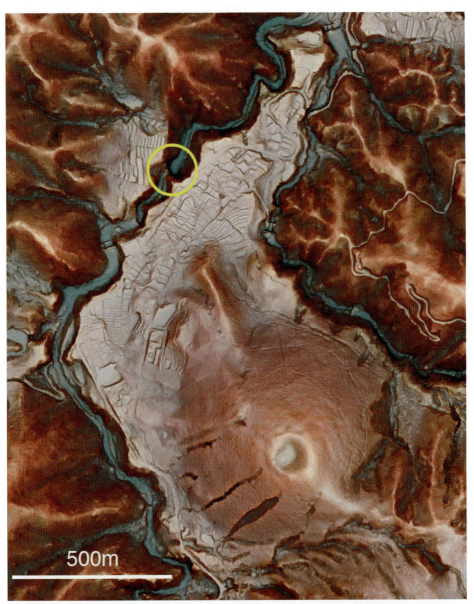

Red Relief Image Map (RRIM) of Hachikuboyama Volcano and the Jorennotaki Falls. Hachikuboyama Volcano (the round hill at the bottom of the map) was created as a result of its eruption approximately 17,000 years ago and the lava flow gushed from the mountain's foot streamed down northward through the valleys. After the eruption, the edge of the lava flow was eroded by the river, which resulted in creating what is now the Jorennotaki Falls (the bump in a yellow circle) (refer to Section 46). The image was provided by the Numazu River and National Highway Office of the Ministry of Land, Infrastructure, Transport and Tourism.

Frontispiece 7

The Jorennotaki Falls runs down onto the lava flow effused from Hachikuboyama Volcano (see Frontispiece 6 and refer to Section 46).

The layer of piles of volcanic ash erupted by Omuroyama Volcano (refer to Section 49).

Frontispiece 8

The port of Heda is located in the northwestern part of the Izu Peninsula. The mountain behind the port was created by Daruma Volcano, which is one of the large terrestrial volcanoes formed after the collision between Izu and the Japanese main island (refer to Sections 29, 30 and 33). The picture was taken by the Izu Peninsula Geopark Promotion Council.

A cliff around Cape Osezaki located at the northwestern edge of the Izu Peninsula. The cliff is heaped with piles of lavas effused from Osezaki Volcano, a large terrestrial volcano (refer to Section 33).

Frontispiece 9

A bird's-eye view from the Tanna Basin to Fuji Volcano. The countryside area at the lower left of the picture is the Tanna Basin. The Tanna Fault, a major active fault on the Izu Peninsula, extends north and south through the middle of this basin to Hakone Volcano at the upper right of the picture (refer to Sections 62 to 64). Fuji Volcano can be seen at the upper left of the picture.

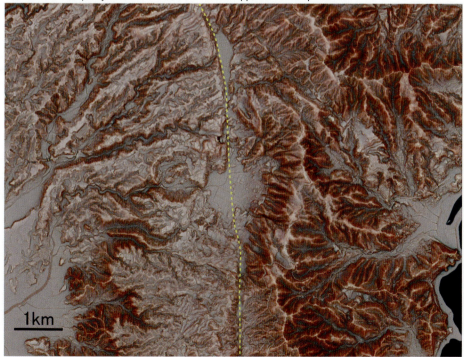

Red Relief Image Map (RRIM) of the Tanna Fault (refer to Sections 62 to 64). Dashed line in the middle of the picture is the Tanna Fault, which extends along the ravines that run north and south. The square-shaped depression at the center of the picture is the Tanna Basin. The image was provided by the Numazu River and National Highway Office of the Ministry of Land, Infrastructure, Transport and Tourism.

Frontispiece 10

A bird's-eye view of the Izu Peninsula seen from the southeastern side of Shimoda. At the front of the picture is the Suzaki Peninsula and the bay on its left is the port of Shimoda. The coast at the right back of the Suzaki Peninsula is the Shirahama Coast. The Suzaki Peninsula is flat and was formed as a result of the uplift of shallows. As evidence for this, you can observe coastal terraces with several steps, wave-cut benches and wave-cut notches in many places of the Suzaki Peninsula (refer to Section 68).

A bird's-eye view of the Izu Peninsula seen from the southeastern side of Yumigahama. The sandy beach at the front of the picture is Yumigahama, which forms the east coast of the Izu Peninsula, and just on the side of the coast is the mouth of the Aonogawa River. The picture clearly shows that the ravines above the Aonogawa River extend to the area near the west coast of the peninsula. This can be considered asymmetrical landforms created by the crustal movement of the Izu Peninsula (refer to Section 69).

Frontispiece 11

The beautiful strata of the Dogashima Coast in western Izu formed by submarine volcanic eruptions (refer to Sections 14 to 15). Beautiful laminations were created on the strata of pumice and volcanic ash (at the upper left of the cliff) under the influence of waves or ocean currents.

The cliff in the Sawada Park to the south of the Dogashima Coast in western Izu. The uppermost layer of lapilli forms strata of subaqueous volcaniclastic flows that ran down the sea floor along with submarine volcanic eruptions (refer to Section 14). The striped strata just below them were created as a result of pumice and volcanic ash having piled up on the seabed under the influence of waves or sea currents. Their undulations show the traces of lower-layer strata liquefied by the weight of subaqueous volcaniclastic flows.

Frontispiece 12

A cliff in the Karano Park to the south of the Dogashima Coast in western Izu. The black angular rocks are hyaloclastic lavas effused from submarine volcanoes that flowed broken by the chilling of seawater (refer to Section 13). The yellow striped strata sandwiched between them were created as a result of pumice and volcanic ash having piled up on the seabed under the influence of waves or sea currents.

You can see strata of subaqueous volcaniclastic flows that drifted down the sea floor along with submarine volcanic eruptions on the cliff of Mt. Hiyoriyama in the Koura area, Minami-Izu Town (refer to Section 14). The large rocks are chilled volcanic bombs (refer to Section 15) and have characteristic fissures (chilled margin), which provides evidence that it touched seawater while they were at a high temperature.

Frontispiece 13

The Senganmon Rock along the coastline on the southern side of Matsuzaki in western Izu. It is one of the "volcanic necks" created as a result of what used to be a submarine volcano having been eroded and its core having appeared on the surface of the earth (refer to Section 17).

"Jakudari (snake-crossing)" dike seen on the coast of Koura in southern Izu. The horizontal striped strata are strata of pumice and volcanic ash that piled up on the sea floor. A clear dike that run across those strata can be seen. The dike was created as a result of magma having filled an underground fissure and having chilled and hardened (refer to Section 17). The dike has columnar joints formed by cooling and contraction. The picture was taken by Yusuke Suzuki.

Frontispiece 14

Strata of volcanic ash turbidite seen on the bank of the Kanogawa River. These strata were created as a result of pumice and volcanic ash having run down the slope of the seabed and having piled up (refer to Section 8).

Pillow lavas seen around the Nishinagawa River, which runs through western Izu. These pillow lavas were created as a result of low-viscosity lavas effused onto the sea floor having become tube-shaped flows by surface tension (refer to Section 4). They are part of the Nishina Group, the oldest (approximately 20 million years ago) strata seen on the Izu Peninsula.

Frontispiece 15

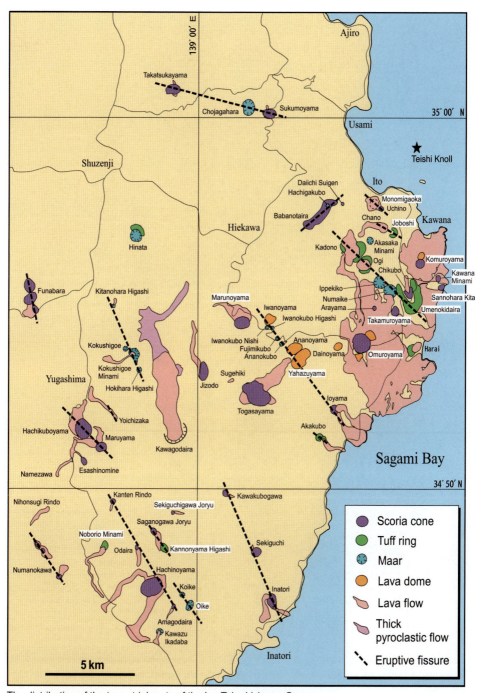

The distribution of the terrestrial parts of the Izu Tobu Volcano Group

Frontispiece 16

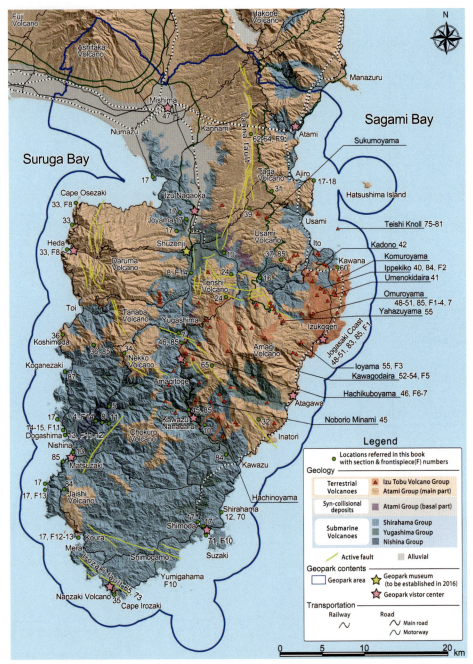

Index map showing the locations of primary geosites of the Izu Peninsula noted in this book. The base map was drawn up by the Izu Peninsula Geopark Council. Three-dimensional geomorphology was drawn from the Fundamental Geospatial Data developed by the Geospatial Information Authority of Japan.

Geohistory of the Izu Peninsula

By MASATO KOYAMA

THE SHIZUOKA SHIMBUN

Geohistory of the Izu Peninsula
By Volcanologist Masato Koyama, translated by Kazuya Hirai

Publisher

The Shizuoka Shimbun

3-1-1 Toro, Suruga-ku, Shizuoka-city, Shizuoka pref., 422-8033, Japan

Copyright © 2015 by Masato Koyama

All rights reserved. No part of this book may be reproduced or utilized in any means, electronic or utilized in any form or by any information storage and retrieval system, without permission in writing from the publisher.

Printed in Japan

ISBN978-4-7838-0550-2

About the Author

Masato Koyama is a professor at the Center for Integrated Research and Education of Natural Hazards (CIREN) of Shizuoka University and at the Faculty of Education of Shizuoka University. Professor Koyama specializes in volcanology and mitigation of earthquake and volcanic disasters. He was born in the city of Hamamatsu, Shizuoka, in 1959. Koyama obtained a doctoral degree in science (geology) from the University of Tokyo. Koyama held posts as a director of the Volcanological Society of Japan and a member of the Committee for Hazard Maps of Fuji Volcano. Currently, he is a member of the Volcano Disaster Management Council for Izu Tobu Volcano Group and the Izu Subcommittee of the Coordinating Committee for Prediction of Volcanic Eruption and also the advisor to the Izu Peninsula Geopark Promotion Council. Koyama has continued geological studies of Izu for more than thirty years since his university graduation studies.

About the Translator

Kazuya Hirai is a professional translator. He is a member of the Japan Association of Translators (JAT). His profile is on http://jat.org/translators/5104/

Cover photo

The coast of Oku Iro in summer. The tableland (Ikenohara) at the front, where day lilies blossom in full glory in the evening, was formed by the eruption of Nanzaki Volcano (refer to Section 35).

Foreword

This book describes Izu Peninsula's distinctive history in which its foundation continued to drift northward, while it was born and grew as a result of submarine eruptions in the sea far south of it, until it finally became its current shape of peninsula after its collision with the Japanese main island. In addition, the book also introduces current ongoing earthquakes and volcanic activities in Izu and on its periphery and discusses how to coexist with the earth with a focus on their forecasts and disaster damage control issues.

What I would like to consistently highlight in this book is that all of familiar landforms and landscapes have their own meanings. If you learn to interpret those meanings on your own, the world you see will appear quite different. If you just acquire the basic knowledge explained in this book, you will be able to discern hidden meanings and backgrounds from landforms and landscapes that you considered just beautiful up until now. Only those who undergo such experience can know the unspeakably amazing intellectual excitement and pleasure it brings.

Those who have learned to make out the meanings of landforms and landscapes can imagine dynamic volcanic activities and crustal movements that created the earth of Izu as if they had directly seen those things. In other words, it means acquiring the ability to discern the particular dangers of natural disasters to particular areas and also the ability to see through the fact that natural disasters will bring great blessings to human society in the long run. That is, it means acquiring the ability to gain a balanced grasp of natural risks and benefits.

An increase in the number of people with those abilities leads to boosting fundamental improvements in the whole society's approach to nature and the quality of the public sense of disaster damage control and will eventually result in building a society that is strong against disasters. In addition, it will no longer be a dream to realize new types of tourism and regional promotion emphasizing the benefits of nature. The perfect example of this is the concept of "geopark" that seeks to preserve valuable landforms and geological heritages and put them to their utmost use.

I sincerely hope that this book's broad readerships of people who love the earth of Izu will contribute to minimizing future natural disasters on the peninsula and that the unknown value and greatness of Izu's nature will also be disseminated to the whole world.

The winter of 2014, Masato Koyama

The author greatly thanks to Prof. Yukihiro Ito, the president of Shizuoka University, because the cost of publication of this book was supported by the president's discretionary expenses.

Contents

Introduction   9

1. Izu came from the South   10
2. The geological history of the Izu Peninsula   12
3. How to measure the geologic age of strata   14

Chapter 1: The period of submarine volcanoes   17

4. The Nishina Group (1): Submarine lava flow   18
5. The Nishina Group (2): Submarine volcanic debris flow   20
6. The Nishina Group (3): Strange characteristics of rock   22
7. The Yugashima Group (1): Strata along the valleys   24
8. The Yugashima Group (2): Turbidity current   26
9. The Yugashima Group (3): Green rocks   28
10. The Yugashima Group (4): Fossils of the South Seas   30
11. The Yugashima Group (5): What happened 10 million years ago?   32
12. The Shirahama Group (1): The Pliocene Coast   34
13. The Shirahama Group (2): The uplift of submarine volcanoes   36
14. The Shirahama Group (3):
    Beautiful strata in Dogashima: Subaqueous volcaniclastic flows   38
15. The Shirahama Group (4): Beautiful strata in Dogashima: Chilled bomb   40
16. The Shirahama Group (5): White cliff   42
17. The Shirahama Group (6): Volcanic neck   44
18. The Shirahama Group (7): First land   46

Chapter 2: The path to a peninsula   49

19. Drilling into the sea floor around Izu (1): Deep-sea drillship   50
20. Drilling into the sea floor around Izu (2): International joint study   52

21. Drilling into the sea floor around Izu (3): The northward drift of plates 54
22. Izu's collision with the Japanese main island (1):
Closed channel: The Ashigara Group 56
23. Izu's collision with the Japanese main island (2):
Closed channel: Landfill and uplift 58
24. Izu's collision with the Japanese main island (3): The last sea 60
25. Izu's collision with the Japanese main island (4): The indentation of Izu 62
26. Diving in the bottom of the Suruga Bay (1): *SHINKAI 2000* 64
27. Diving in the bottom of the Suruga Bay (2): The Dive 579 launched 66
28. Diving in the bottom of the Suruga Bay (3): Sinking sea floor 68

Chapter 3: The period of terrestrial large volcanoes 71

29. A long range of multiple volcanoes: Polygenetic volcano 72
30. A long range of multiple volcanoes: Lost summit 74
31. Yugawara Volcano, Taga Volcano and Usami Volcano 76
32. Amagi Volcano and Tenshi Volcano 78
33. Daruma Volcano, Ita Volcano and Osezaki Volcano 80
34. Tanaba Volcano, Nekko Volcano and Chokuro Volcano 82
35. Jaishi Volcano and Nanzaki Volcano 84
36. Volcano that created glass 86
37. Obsidian in Izu 88

Chapter 4: The period of the Izu Tobu Volcano Group 91

38. A group of small volcanoes 92
39. Pumice and volcanic ash from Hakone 94
40. Lake Ippekiko and Numaike 96
41. Umenokidaira 98
42. Kadono and Ogi 100
43. What a volcano chain means 102
44. The largest eruption of Hakone Volcano 104
45. A volcano that created the Kawazu Nanadaru Falls 106
46. Hachikuboyama and the Jorennotaki Falls 108

47. The eruption of Fuji Volcano and Izu 110

48. Omuroyama (1): Scoria cone and crater 112

49. Omuroyama (2): Lava outlet 114

50. Omuroyama (3): A dammed lake and lava coast 116

51. Omuroyama (4): Pothole and scoria raft 118

52. Kawagodaira (1): Lava flow and pyroclastic flow 120

53. Kawagodaira (2): The entire process of eruption 122

54. Kawagodaira (3): Horrible eruption 124

55. Iwanoyama- Ioyama volcano chain 126

56. A summary of eruptive history (1): Locations of eruption and magma types 128

57. A summary of eruptive history (2): Serious future 130

58. A summary of eruptive history (3): Doughnut-shaped structure 132

Chapter 5: The living earth of Izu 135
(earthquakes and crustal movement)

59. The current situaiton surrounding Izu in terms of earth science 136

60. The Tokai Earthquake, the Nankai Earthquake and the Kanto Earthquake 138

61. The West Kanagawa Prefecture Earthquake 140

62. The Tanna Fault (1): Disconnected ravines 142

63. The Tanna Fault (2): The Kita Izu Earthquake and trench excavation 144

64. The Tanna Fault (3): The past and the future 146

65. A country of active faults 148

66. The mystery of tectonic rotation: Unusual fault distribution 150

67. The mystery of tectonic rotation: The mechanism of rotation 152

68. Coastal landforms tell 154

69. A peninsula inclined to the west 156

Chapter 6: The living earth of Izu (magmatism) 159

70. The genealogy of volcano deities 160

71. The illusion of eruption 162

72. Earthquake swarms in 1930 164

73. From 1930 to 1978 166

74. The Off-Ito Submarine Eruption of 1989 (1): The process to eruption   168

75. The Off-Ito Submarine Eruption of 1989 (2): Volcanic tremor and eruption   170

76. The Off-Ito Submarine Eruption (3): The shock of eruption   172

77. The Off-Ito Submarine Eruption of 1989 (4): Given time for preparation   174

Chapter 7: Coexisting with the earth   177

78. Predicting magmatism (1): Success in predicting its start   178

79. Predicting magmatism (2): Predicting scale and end   180

80. Predicting magmatism (3):
Volcano monitoring and disaster damage control system   182

81. Predicting magmatism (4): Magmatic activity scenario   184

82. Learning volcanoes   186

83. The blessings of volcano (1): Volcanoes create land   188

84. The blessings of volcano (2): Volcanoes create water sources   190

85. The blessings of volcano (3): Volcanoes produce stone and tourist resources   192

86. The dream of geopark (1): Mother earth   194

87. The dream of geopark (2): The assets of geopark   196

88. The dream of geopark (3): The current situation of Izu Peninsula Geopark   198

Introduction

1. Izu came from the South

In February 1931, Yasunari Kawabata, who was awarded the Nobel Prize in Literature in 1968, said, "Izu is a model of the South." He gave such a figurative expression to the unique features of the scenic beauty and vegetation of Izu that are different from those of the Japanese main island. However, currently, over 80 years afterward, the marvelous fact that his language was actually not a metaphor but that the earth of Izu originally "derived from the South" in a real sense of the word has emerged. The foundation of the Izu Peninsula was originally created in low latitudes and reached where it is today after a long journey.

The Izu Peninsula is located in a geologically unique place. The Earth surface is covered by bedrocks called plate with a thickness of dozens of kilometers to one hundred kilometers. Those plates vary in size and slowly move in different directions along with the convection inside of the Earth. Four plates are piled up on each other around Japan. The Izu Peninsula is situated at the northern end of the Philippine Sea Plate. The Philippine Sea Plate is subducting under the Amurian (Eurasian) and the Okhotsk (North American) Plates, on which the Japanese main island sits. The Pacific Plate is slowly subducting under the Okhotsk and Philippine Sea Plates.

There is a protuberance of volcanic islands and submarine volcanic rows called the Izu-Bonin Arc along the eastern end of the Philippine Sea Plate. If a plate that subducted into the inside of the Earth reaches an underground point approximately one hundred kilometers, dehydration occurs, which causes the melting point to drop and a huge amount of magma to come out. Many volcanoes were born and grew on the Japanese archipelago and the Izu-Bonin Arc when magma created by the subduction of the Pacific Plate floated to Earth's surface. Those volcanoes include the Izu Peninsula, the Izu Shichito Islands, Tori-shima, Iwo Jima and many volcanic islands and submarine volcanoes around them. Most of the earth that constitutes the Izu Peninsula is formed by volcanic products from many volcanoes that used to exist on land and at the bottom of the sea.

The Philippine Sea Plate, on which the Izu Peninsula sits, is slowly drifting northwestward toward the Japanese main island at a few centimeters per year. Although this pace seems to be negligible, the plate will have moved dozens of kilometers in one million years. The base of the Izu Peninsula was formed approximately 40 million years ago, which means that at that time what is now the Izu Peninsula was located more than one thousand kilometers south of where it is

Introduction

today. That location is where Iwo Jima is today.

As these above-mentioned descriptions clearly show, the whole Izu Peninsula used to be a volcanic island (partly a submarine volcano) floating in the South Seas. It was 600,000 years ago that Izu collided with the Japanese main island and got peninsula-shaped.

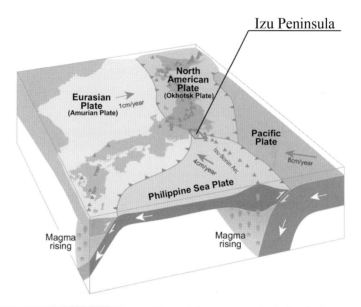

▲ Four plates are complicatedly piled up on each other around the Japanese archipelago. The Izu Peninsula is located at the northern extremity of the Philippine Sea Plate. The triangle marks show active volcanoes. The arrow marks on the Earth surface and the figures just beside them show the direction and speed of plate motion on the basis of Northeast Japan (the North American Plate).

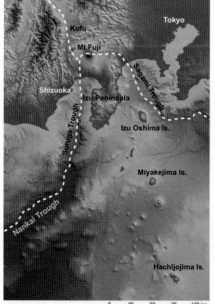

◀ The geomorphology around the Izu Peninsula and plate boundaries (dashed line). This map was made from the Fundamental Geospatial Data by the Geospatial Information Authority of Japan (terrestrial part) and the 500m Mesh Depth-sounding Data by the Japan Oceanographic Data Center (marine part).

2. The geological history of the Izu Peninsula

Submarine and terrestrial volcanic eruptions spanning 40 million years created the earth of the Izu Peninsula. The strata of approximately 20 million years in half can be seen on the current surface of the peninsula and the strata of another 20 million years are buried underground. Therefore, a scrupulous examination of the strata in mountains and on coastal cliffs on the Izu Peninsula enables you to track its history from approximately 20 million years ago to the present.

As a result of these studies, the geological history of the Izu Peninsula is segmented into the following five periods.

(1) The period of submarine volcanism in the deep sea (20 million to 10 million years ago)

Around this period, Izu formed a submarine volcanic group that was hundreds of kilometers south of the Japanese main island. The strata created by seabed piles of lava, volcanic blocks/lapilli and volcanic ash ejected from those submarine volcanoes are called the Nishina Group and the Yugashima Group in ascending order.

(2) The period of submarine volcanism in the shallow sea (10 million to two million years ago)

There were some volcanoes that appeared on the sea surface and became volcanic islands because the whole of Izu became a shallow sea. The strata of lava, volcanic blocks/lapilli and volcanic ash that erupted around this period are called the Shirahama Group.

(3) The start of Izu's collision with the Japanese main island and the appearance of terrestrial areas (two million to one million years ago)

During this period, Izu collided with the Japanese main island and was becoming part of the main island. Most of Izu became land by this collision for the first time in history and afterward, all volcanoes erupted on land. The deposits in and after this period are called the Atami Group.

(4) The period of large terrestrial volcanoes (one million to 200,000 years ago)

By 600,000 years ago, Izu had got the shape of a peninsula that protruded from the Japanese main island and the original form of what is today the Izu Peninsula was created. Eruptions occurred everywhere on the Izu Peninsula that had become land and large volcanoes, such as Amagi Volcano and Daruma Volcano, were born.

(5) The period of the Izu Tobu Volcano Group (200,000 years ago to the present)

Introduction

By around 200,000 years ago, all volcanoes except for Hakone Volcano had stopped erupting while the Izu Tobu Volcano Group started to erupt. The last eruption of this volcano group was the submarine eruption of Teishi Knoll off the coast of Ito in July 1989.

In the following sections, I will take a close look at each period, noting specific examples.

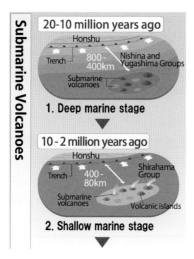

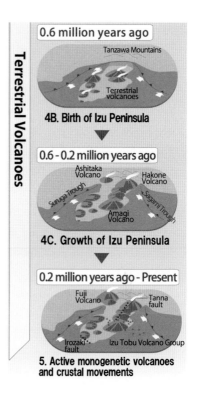

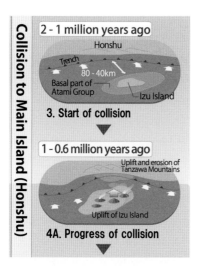

The geological history of the Izu Peninsula. The illustrations depict a rough image of Izu and its periphery during the periods from (1) to (5) mentioned above (the fourth period is divided into 4A, 4B and 4C).

3. How to measure the geologic age of strata

When experts speak about strata, they often mention an overwhelmingly old geologic age, saying, "This stratum is hundreds of thousands of years old." I speculate that some readers of this book wonder about this kind of thing. How do they measure a mind-boggling old age compared with the life span of people?

Multiple methods to measure the geologic age of strata are developed. These methods include the one to measure the radiometric age of rocks contained in strata, the one to study the type of fossils contained in strata and the one to measure paleomagnetism recorded in strata. There are more other methods, but in this context, I will give a detailed explanation about these three methods that played an effective role in measuring the geologic age of Izu's strata.

Firstly, the radiometric age measurement method can estimate the passage of time after the formation of rocks by studying the amount of radioactive elements contained in rocks and the amount of elements formed through the disintegration of those radioactive elements. This method utilizes a mechanism in which radioactive elements disintegrate at a pace peculiar to them. However, this method can only be applied to elements concentrating in particular volcanic rocks, such as potassium, and carbon contained in plant fossils. The applicable age range is also limited by element type.

Secondly, the biochronological method using fossils becomes available by conducting thorough studies of when a particular species of animals or plants appeared and became extinct and by creating a time scale. Of course, the supporting role of the radiometric age measurement method and other methods is necessary for the creation of such time scale. As long as you can create a biochronological time scale, you have only to find key fossils and fossil combinations from the inside of strata. For particular microorganisms, such as calcareous nannoplankton, planktonic foraminifera and radiolaria, time scales on their fossils have been fully developed and are very useful.

It is known that for the current geomagnetic field of the Earth in which the N-pole points to the north, the S-pole used to occasionally point to the north. The former case is called normal polarity and the latter reversed polarity. These directions of geomagnetic field are recorded in the magnetic minerals of rocks and these records form a geomagnetic polarity time scale that provides detailed information about the change between normal and reversed polarities. That is why measuring the remanent

Introduction

magnetization of a rock enables you to get a broad range of information about when the rock was formed.

A device to measure the remanent magnetization of rocks. The cylinder is a sextuple magnetic shield to intercept the current geomagnetic field.

A tree buried under the volcanic products ejected from the Izu Tobu Volcano Group. The tree appears black because it has been carbonized. The radiometric age measurement method using carbon can be applied to them

Chapter 1
The period of submarine volcanoes

4. The Nishina Group (1): Submarine lava flow

The Nishinagawa River flows into a seashore two kilometers south of Dogashima, which is known for its scenic beauty along the west coast of the Izu Peninsula. The river forms almost the deepest and longest ravine in western Izu. If you drive up Route 59 more than ten kilometers along the river from its mouth, you will come to the mountainous small community of Miyagahara, which is the most distant from the sea in Nishi-Izu Town. This area is just 300 meters above sea level and is surrounded by steep mountains nearly 1,000 meters high. The Nishina Group, which is the oldest strata on the Izu Peninsula, is distributed along the valleys stretching from the middle course of the Nishinagawa River to its downstream (See Frontispiece 16).

A large part of the Nishina Group is taken up by volcanic deposits, including lava flows from submarine eruption and subaqueous volcaniclastic flows created as a result of piles of volcanic products collapsing and running down the seabed slopes. Many of those lava flows have a special shape called pillow lava. If low-viscosity lavas flow down the sea floor, their surface tension in seawater create partly constricted tube-shaped flows just like a sausage before cut apart. Such lava is called pillow lava because this "sausage" looks like a pillow.

The occasional lava flows on the seabed of the Kilauea Volcano on the Island of Hawaii are globally known as a major example of pillow lava, but pillow lavas can also be seen in a place where what used to be the seabed was uplifted and became land. In Japan, the pillow lavas on the Okuzure Coast between the city of Shizuoka and the city of Yaizu, Shizuoka prefecture, are particularly well-known for their broad-range distribution and clear shape of pillow, but there are also many examples in other areas.

Pillow lavas of the Nishina Group that can be recognized relatively easily even by laymen are distributed on the cliff along the woodland path in the Ishiki area, Nishi-Izu Town, around four kilometers northeast of the mouth of the Nishinagawa River (see Frontispiece 15 [below]). However, the rocks are approximately 20 million years old and the shapes of the pillows are vague due to the alteration and weathering of the lavas in the passage of time. That is why you need to get your eyes used to the lava shapes to fully recognize them as pillow lavas.

The period of submarine volcanoes

The Nishina Group's pillow lavas effused by submarine eruption (the Ishiki area in Nishi-Izu Town). The cross sections of each pillow can be vaguely recognized. These pillows were created as a result of the cooling and solidification of tube-shaped lavas. Because the lavas flowed down to this place many times, multiple pillows were piled up on each other. The picture was taken by Yusuke Suzuki.

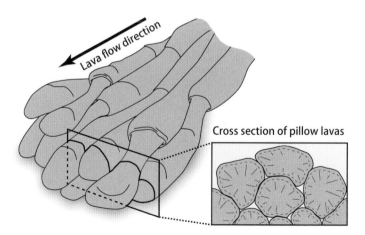

The illustration depicts pillow lava flows. The illustration was drawn up by the Izu Peninsula Geopark Promotion Council.

5. The Nishina Group (2): Submarine volcanic debris flow

Just like many other strata in Izu, the Nishina Group, which is the oldest strata on the Izu Peninsula, was created as the result of what used to be a submarine volcano uplifting and appearing on land. As evidence for this, you can observe what is called subaqueous volcaniclastic flow in addition to the pillow lava noted in the previous section.

Terrestrial debris flows are a strong flood of streams containing a mixture of water, soil and stone that collapsed in a heavy rain and the like. Similar phenomena are known to occur at the bottom of the sea and a lake and these are altogether called subaqueous volcaniclastic flow. It is thought that subaqueous volcaniclastic flows are caused by terrestrial landslides and debris flows pouring into the water or a massive amount of volcanic ash and blocks collapsing underwater in case of a major earthquake and a volcanic eruption.

Terrestrial debris flows result in producing a pile of sediments containing mixtures of rocks, sands and mud with different sizes. In the case of subaqueous volcaniclastic flows, however, water causes heavy and large rocks to sink first and light and small mud and sand to pile up later. Because of this, strata created by subaqueous volcaniclastic flows contain many large and heavy rocks in the lower layer and a cluster of light mud and sand in the upper layer. Piles of strata with these characteristics can be seen on the riverside and in mountains from around the Ishiki area in Nishi-Izu Town, four kilometers northeast of the mouth of the Nishinagawa River, to around the Deai area in the same town, which is located six kilometers northeast of the Ishiki area. Some rock fragments contained in these strata have chilled structures that were formed by fresh hot lavas meeting seawater, which presents a vivid image of submarine volcanic eruptions that occurred in the distant past.

A thin layer of mud that slowly piled up after debris flows had stopped can sometimes be seen in the top upper part of strata created by subaqueous volcaniclastic flows. Fossils of microorganisms (calcareous nannoplankton) that used to live in the sea were discovered from inside the mud and their characteristics made it possible to identify the geologic age of the Nishina Group (the strata containing fossils are approximately 17 million years old and old lavas lying in the lower layer are probably 20 million years old).

The period of submarine volcanoes

Strata of subaqueous volcaniclastic flows formed by submarine volcanic eruptions (the Deai area in Nishi-Izu Town). A layer that is primarily comprised of light and small debris. Although the strata are covered by concrete today, the fossils of the oldest plankton on the Izu Peninsula (17 million years old) have been discovered from inside this cliff along a prefectural road.

Strata of subaqueous volcaniclastic flows formed by submarine volcanic eruptions (the Deai area in Nishi-Izu Town). A layer that is primarily comprised of heavy and large rocks.

6. The Nishina Group (3): Strange characteristics of rock

As mentioned above, the Nishina Group was created as the result of what used to be a submarine volcano uplifting and appearing on land. Similar submarine volcanic products can be seen in large numbers in strata in other ages. However, the characteristics of volcanic rocks contained in the Nishina Group are different from those of volcanic rocks contained in other strata in Izu.

First of all, volcanic rocks contained in the Nishina Group are mostly aphyric basalt. Phenocrysts are large crystals contained in volcanic rocks in the form of spots and being aphyric means including few phenocrysts. Most volcanic rocks contain phenocrysts. Of course, some volcanic rocks are aphyric, but aphyric volcanic rocks contained in the Nishina Group are noticeably extraordinary in width and thickness.

For a second characteristic in terms of naked eyes, volcanic rocks contained in the Nishina Group include a large number of dark green tiny spots different from crystals on their greenish gray surface. These small spots are the remnants of bubbles contained in old rocks. When magmas erupt, gases melted into them evaporate and bubbles form in large numbers. After these bubbles cool down and harden, they stay within the rocks. Bubbles in fresh volcanic rocks are hollow. However, if they undergo geothermal alteration by hot spring water and the like, altered minerals are produced within the bubbles. In the case of the Nishina Group, the geothermal alteration of underlying rocks themselves progressed and the whole rocks got greenish. Inside those rocks are many dark green spots that resulted from altered minerals filling in what used to be hollows. This characteristic is so unique that when a rock of this kind is found, its origin can be identified as the Nishina Group. In Izu, this rock is sometimes used for the bathtub and paving stones of the washing space in a hot spring inn.

Volcanic rocks contained in the Nishina Group are also unique in terms of chemical composition. An examination of the chemical composition of a rock enables you to estimate where on Earth the rock was formed to some extent. Volcanic rocks contained in the Nishina Group are richer in magnesium and the like than other volcanic rocks on the Izu Peninsula and the Japanese archipelago. It is considered that those rocks did not originate from magmas formed by normal plate subduction. The Nishina Group seems to show that when the Shikoku Basin stretching to the south of the Japanese archipelago was formed by plate rifting, part of volcanic activities in Izu was also influenced by that.

The period of submarine volcanoes

A ravine cut by the Nishinagawa River running into the west coast of Izu.

A volcanic rock often seen in the Nishina Group. A large number of dark green tiny spots on its green and gray surface. (the Nishina area in Nishi-Izu Town)

7. The Yugashima Group (1): Strata along the valleys

The Yugashima Spa is a renowned hot spring situated along the valley up the Kanogawa River almost in the middle of the Izu Peninsula and is also known as a place noted in connection with Yasunari Kawabata, a Nobel Literature Award-winning writer. The Yugashima Group is known as strata with the same name. The strata are proximately 15 million to 10 million years old. The Yugashima Group is the second oldest strata on the Izu Peninsula after the Nishina Group (20 million to 16 million years old), which I explained in the previous sections.

General rules and the International Stratigraphic Guideline stipulate that the names of strata must be given the names of areas (type locality) in which they are typically distributed. Strictly speaking, Group is a general name for multiple strata, but is usually given an area name. Because the Yugashima Group is distributed across the whole Izu Peninsula, the area around Yugashima in which the most typical examples of strata can be observed was chosen and the strata were given that name.

As the illustrated distribution of the Yugashima Group shows, the group is mainly distributed along the valleys beside large rivers on the Izu Peninsula and is not distributed around high-altitude ridges and mountain tops (see Frontispiece 16). This is because new strata (the Shirahama Group and the Atami Group) sitting on the Yugashima Group occupy high-altitude mountainous areas.

It is many hot springs on the Izu Peninsula that are concentrated along the valleys just like the Yugashima Group. Most of these hot springs were created after rainwater and seawater had soaked deeply into the underground and had spouted to the Earth surface, heated at a spot with high geothermal heat. Therefore, there are many cases in which hot springs spout along the valleys and near seashores rather than in places at a high altitude. Of course, there are some exceptions, but it is for this reason that many hot springs spout from the Yugashima Group.

Various minerals precipitate in hot spring water. To put it simply, many old gold mines on the Izu Peninsula where people used to dig out gold are precipitations of minerals including gold, from hot spring water in cracks in the rocks. This is why the Yugashima Group is also known as strata where there are many hidden gold mines.

The period of submarine volcanoes

An example of the Yugashima Group's strata on the riverside around the Yugashima Spa

Another example of the Yugashima Group's strata on the riverside around the Yugashima Spa

8. The Yugashima Group (2): Turbidity current

The most remarkable characteristic of the Yugashima Group, the second oldest strata on the Izu Peninsula, is many clear stripes of the strata. Probably, general people may take it for granted that the stripes of strata are clear. However, the "stratum" in a geological sense includes many strata without stripe (called stratification). Stripes of strata are easily formed where rock particules are fine. In particular, clear stripes are often created by water flows in the case of strata containing clastic particles smaller than the size of sand grains.

Because the Yugashima Group includes many strata containing such minute clastic particles, clear stripes can be observed. Many of these clastic particles are volcanic ash ejected from submarine volcanoes everywhere. Such volcanic ash piles up on the sea floor around volcanoes, but they break up, triggered by subsequent eruptions or earthquakes, run deeper down on the seabed's slope in groups. This phenomenon is called turbidity current. At the beginning of the famous movie *Japan Sinks* (1973), there is a scene in which the deep-sea submarine that the protagonist is aboard encounters big turbidity currents. Similar phenomena frequently occur near a submarine volcano.

If there is a basin beside a submarine volcano, the turbidity current repeatedly runs down toward there. The strata containing mud and sand that piled up under the influence of the turbidity current are called turbidite and if many of the mud and sand are volcanic ash, the strata are called ash turbidite. When the turbidity current runs, heavy and large particles sink below and light and small particles float above. That is why if you closely observe a turbidite, you can find the downsizing of particles from below to above. Many of the strata with clear stripes included in the Yugashima Group are ash turbidites and can be well observed on the cliff and the riverside typically in the Kadono area, the city of Izu, and in the Sakurada area, Matsuzaki Town (see Frontispiece 14 [above]).

The period of submarine volcanoes

Ash turbidite of the Yugashima Group (on a cliff along the Kanogawa River in the Hinata area, the city of Izu) Many stripes (stratification) can be seen in the strata. The close-up view of the cliff in Frontispiece 14 [above].

Ash turbidite of the Yugashima Group. An example seen in the eastern part of the Ishiki area, Nishi-Izu Town.

9. The Yugashima Group (3): Green rocks

Many of the rocks forming the Yugashima Group underwent significant geothermal alterations under the influence of geothermal heat and hot spring water. As a result of such geothermal alterations, greenish minerals were created within the rocks, which led the whole rocks to get a touch of green color as well. Of course, there are some fairly fresh parts and some strata of the Shirahama Group, which is younger than the Yugashima Group, underwent considerable geothermal alterations. Therefore, it is impossible to identify rocks that are part of the Yugashima Group only from the degree of geothermal alterations. That is why it is mistaken to simplistically think that if the rocks have altered into green, the strata belong to the Yugashima Group, as was often said years ago.

Although these points should be noted, roughly speaking, the Yugashima Group has many greenish rocks. As mentioned in the previous section, the Yugashima Group includes many turbidites formed by submarine flows called turbidity current. However, some parts are made up of lava flows. Turbidity currents and lava flows are terms of classification by the formation process of strata. Primary substances that form the strata of the Yugashima Group are volcanic ash, volcanic lapilli and volcanic blocks. The volcanic lapilli contain pumice and scoria (dark-colored pumice, such as black and blackish brown). The substance of pumice and scoria is volcanic glass that contains many bubbles. These many types of rocks are all greenish because of their alterations under the influence of geothermal heat and water. Interestingly, the geological community calls altered rocks "rotten rocks."

Volcanic glass is particularly susceptible to geological alterations and can sometimes turn into patches of altered green minerals that are quite different from what they originally were. In such case, they become rocks that appear to include many green spots. Some rocks of this kind are named "Izu Young Grass Stone" and have long been used as building materials for a stone fence and a bathtub. They have recently been sold as moisture absorbent and deodorant at souvenir shops.

The period of submarine volcanoes

Altered rocks of the Yugashima Group. They contain rock fragments of many different colors, but are slightly greenish as a whole.

"Izu Young Grass Stone" sold at a souvenir shop in Izu

10. The Yugashima Group (4): Fossils of the South Seas

Some strata of the Yugashima Group contain a large amount of fossils, although such case is rare. There are two major fossiliferous strata: the Nashimoto Limestone (Nashimoto in Kawazu Town) and the Shimoshiraiwa Calcareous Sandstone (Shimoshiraiwa in the city of Izu).

Both the Nashimoto Limestone and the Shimoshiraiwa Calcareous Sandstone are not much altered and many experts have the wrong recognition that they are part of the Shirahama Group, which is younger than the Yugashima Group. However, a study of the distribution and structure of the whole strata demonstrated that the two strata belong to the Yugashima Group. As noted in the previous section, the degree of alterations cannot be a key criterion for identifying strata.

The Nashimoto Limestone can be seen on the cliff on the southern side of the Okuharagawa River, which is a branch stream of the Kawazugawa River near the Kawazu Nanadaru Loop Bridge. The Nashimoto Limestone is slightly reddish white or yellowish white hard rocks with parallel stratification. The limestone contains the fossils of shells, coral and larger foraminifera and planktonic fossils approximately 14 million years old were also discovered.

The Shimoshiraiwa Calcareous Sandstone can be seen in a hilly area along the Omigawa River in the city of Izu. The sandstone on the northern side of the Omigawa River is well known, but it is also distributed along the street on the southern side of the river and in a mountain. As the name shows, the Shimoshiraiwa Calcareous Sandstone is strata containing white calcareous coarse sands. These sands are easy to break and you can easily take fossils from the strata. The strata contain a large number of various fossils and convex lens-shaped *Lepidocyclina* (larger foraminifera) a few millimeters in diameter is a prefecture-designated special natural monument. These fossils are approximately 11 million years old.

Both the Nashimoto Limestone and the Shimoshiraiwa Calcareous Sandstone are sandwiched between volcanic products created by submarine volcanic eruptions and show that an inhabitable environment for many creatures were secured just for some time during an interval of repetitive volcanic activities. Clearly, both fossils contain subtropical materials compared with fossils in other parts of the country in the same age. This is considered to provide evidence that the earth of Izu used to be located in the South Seas and later reached the area around Japan by plate motion.

The period of submarine volcanoes

The Shimoshiraiwa Calcareous Sandstone in the Shimoshiraiwa area, the city of Izu. The whole strata dipped at an angle of 30 degrees to the east under the influence of subsequent crustal movements. These strata are a Shizuoka prefectural government-designated natural monument.

The Nashimoto Limestone in the Nashimoto area, Kawazu Town

11. The Yugashima Group (5):
What happened 10 million years ago?

Strata pile up almost horizontally and the upper segment is younger. However, the whole strata sometimes dip by crustal movement in later years and are also broken apart by faults. Many of the strata that are part of the Yugashima Group were broken apart into small parts by faults and largely dipped (see Frontispiece 14 (above)). However, the Shirahama Group that is superposed on the Yugashima Group is nearly horizontal with some exceptions.

The photo shows a form of typical relationships between the Yugashima Group and the Shirahama Group. This is a case seen in a mountain in Nishi-Izu Town, but can be said to be an example that represents the whole of Izu. The Yugashima Group's strata of ash turbidite (refer to Section 8) dipped at an angle of 60 degrees. Tuffaceous sandstone, limestone and volcanic breccia (Shirahama Group) piled up almost horizontally and cut into the strata of ash turbidite. More gently inclined strata cut into a dipped stratum and piled up onto it. This relationship between strata is called unconformity and the plane on which those strata meet is called unconformity boundary.

The observed unconformity between the Yugashima Group and the Shirahama Group throughout the Izu area means that the whole of Izu was affected by huge crustal movement during the period just before the pileup of the Shirahama Group. In this period, that the Yugashima Group was folded and cut apart by faults. This occurred approximately 10 million years ago.

The causes and mechanisms of this major phenomenon still leaves many unclear factors, but some think that the same thing occurred in a wide stretch of areas from the Izu Shichito Islands to the Oagasawara Islands beyond Izu. In the meantime, a comparison of the chemical components of volcanic rocks contained in the Yugashima Group and the Shirahama Group shows that the components contained in the Yugashima Group involve more of magma that erupted in a place closer to a trench. This suggests that the crustal movements that occurred 10 million years ago were enormous enough to influence the distance between the Izu and Ogasawara trenches and Izu. Future studies are entrusted with the task of finding the answer to the mystery of what happened at that time.

The period of submarine volcanoes

The Hozoin Limestone directly above the unconformity boundary. Sandstones containing volcanic ash cover it. The Ishiki area in Nishi-Izu Town. The picture was taken by Yusuke Suzuki.

12. The Shirahama Group (1): The Pliocene Coast

In *Night on the Galactic Railroad*, one of Kenji Miyazawa's novels, the protagonist Giovanni and his friend Campanella drop in at a fossil excavation site near a station during their travel by galactic railroad train. Although this place is located at a corner of the universe, somehow, it is called the Pliocene Coast. It is a fantastic scene in the novel. In fact, on the Izu Peninsula, there is a place whose image is quite similar to that of the Pliocene Coast—the Shirahama Coast in the city of Shimoda. (See Frontispiece 10 [above].)

Fundamentally, Pliocene means the Pliocene Epoch (the period from approximately five million to 2.6 million years ago) of the Neogene Period on a geologic timescale. The strata on the cliff on the northern side of a beach on the Shirahama Coast are formed by tuffaceous sandstones that were created as the result of pumice and volcanic ash piling in the sea. A close look at the sandstones shows that they contain many fossils of shells, coral and sea urchin. These fossils are the remnants of creatures in the Pliocene Epoch.

The Shirahama Coast on which you can see vivid strata containing a large amount of fossils in the Pliocene Epoch can be said to be exactly the real Pliocene Coast. The characteristics of the fossils and strata show that the place used to be a shallow sea. There may have been a coast nearby at that time. The Shirahama Coast is a rare place where a Pliocene Epoch coast and a contemporary coast happened to meet beyond the walls of millions of years.

Tuffaceous sandstones on the Shirahama Coast are part of the Shirahama Group. Fundamentally, the Shirahama Group was named after the Shirahama Coast. The Shirahama Group is the third oldest strata on the Izu Peninsula after the Nishina Group and the Yugashima Group, as mentioned in the previous sections, and is broadly distributed on the south coast of the peninsula and in mountains (see Frontispiece 16). The group ranges not only in the Pliocene Epoch but also approximately 10 million to two million years ago. Most of the Shirahama Group is formed by submarine volcanic products and strata created by a pile of debris on the nearby shallow ocean floor.

The period of submarine volcanoes

A bird's-eye view of the Shirahama Coast in the city of Shimoda. On the cliff in the middle are tuffaceous sandstones of the Shirahama Group.

Tuffaceous sandstones on the cliff on the Shirahama Coast. The sandstones contain a large amount of fossils of creatures that used to inhabit the shallow sea.

13. The Shirahama Group (2):
The uplift of submarine volcanoes

As mentioned in the previous section, the Shirahama Group is strata created approximately 10 million to two million years ago and is widely distributed across the Izu Peninsula. The Shirahama Group is strata formed by submarine volcanic products and a pile of debris on the nearby shallow seabed. I wrote "submarine volcanic products" as if I had actually seen those eruptions, but no one has ever actually seen submarine volcanic eruption.

The reason why particular strata can be considered the product of submarine volcanic eruption is that there is a set of two concrete discoveries: (1) evidence that the strata were formed on the sea floor; and (2) evidence that when they were formed they were at a high temperature. For the former, the key element is notable characteristics, such as stripes created by waves and sea currents, and the fossils of creatures in the sea. As for the latter, the most significant factor is lava flows and volcanic bombs, which were chilled by seawater.

When lava flows run down the seabed, they can take the characteristic form of pillow lavas (Section 4) and also take the form of hyaloclastic lavas in some cases. When hot lava flows effused on the seabed come into contact with cold seawater and are chilled, they break into pieces by thermal distortion and sometimes turn into a mass of angular rock fragments and rock blocks. This is hyaloclastic lava (see Frontispiece 12 [above]). Volcanic bombs ejected from the crater of a submarine volcano in a hot form can also get characteristic cracks by thermal distortion and glassy shells on their surface. A volcanic bomb with these characteristics is called chilled bomb (see Frontispiece 12 [below]).

If you closely observe the Shirahama Group, you can find evidence of submarine volcanic eruptions, such as hyaloclastic lava and chilled bomb, everywhere. This evidence drives home to people the fact that Izu, which used to be an aggregation of submarine volcanoes, was uplifted and became land later.

The period of submarine volcanoes

An example of chilled bomb contained in the Shirahama Group. The chilled bomb got radial cracks because it was rapidly deprived of heat from all directions. The picture was taken around Bentenjima in Matsuzaki Town.

An example of hyaloclastic lava contained in the Shirahama Group. Rocks with the same petrologic characters were broken like a jigsaw puzzle. The picture was taken around the port of Nishina in Nishi-Izu Town.

14. The Shirahama Group (3):
Beautiful strata in Dogashima: Subaqueous volcaniclastic flows

On the Izu Peninsula, which has a mountainous terrain, many coasts have cliffs. Some cliffs are even more than 200 meters in height. Those cliffs mostly involve various strata on the surface. Of course, strata can be seen on the cliffs along roads and rivers. However, coastal cliffs are washed by waves, wind and rain, which makes them a perfect place to constantly observe fresh surfaces. One of such places is the Dogashima Coast (Nishi-Izu Town), which is a famous scenic spot on the west coast of the Izu Peninsula (see Frontispiece 11 [above]).

A significant factor that makes the Dogashima Coast a major tourist destination on the west coast of the Izu Peninsula is probably its strikingly beautiful strata on the cliffs. In particular, famous white strata around Tensodo are formed by pumiceous tuff created by a pile of white pumice and volcanic ash on the sea floor. On the surface of the strata are beautiful stripes just like ripples in a dune. This stripe is called cross lamination and was created as the result of rock fragments drifting and being rearranged by waves and sea currents.

Please pay attention to the lower part of the same cliff (the part below the dotted line in the photo below). You can see that the size of rock fragments and rock blocks is gradually getting smaller toward the upper part of the cliff from around the surface of the sea. This is a characteristic of subaqueous volcaniclastic flows, which I explained in Section 5. A flood of flows with a mixture of soil, stone and water is debris flow and this kind of thing that occurred at the bottom of the sea and a lake is called subaqueous volcaniclastic flow. When subaqueous volcaniclastic flows run, heavy and large rocks sink first and light and small mud and sand pile up later. This is why the size of rocks is smaller in the upper part of strata.

By analyzing the thermal history of rocks based on magnetic measurement, evidence was obtained showing that the inside of large rock blocks contained in Dogashima's subaqueous volcaniclastic flows was from 450 to 500 degrees Celsius when they first piled up on the sea floor. This demonstrates that the direct cause of the formation of the strata was the eruption of a submarine volcano.

The period of submarine volcanoes

A bird's-eye view of the coastline from Dogashima in Nishi-Izu Town to the port of Nishina

Strata of subaqueous volcaniclastic flows on the cliff of the Dogashima Coast. Rock fragments get smaller in size from below to above. At the very top is pumiceous tuff with cross lamination. Refer to the text of this and next sections for explanations about the dotted line and the arrow.

15. The Shirahama Group (4):
Beautiful strata in Dogashima: Chilled bomb

The part below the dotted line in the photo specified in the previous section is strata of subaqueous volcaniclastic flows, which I explained in the previous section, and the part above the dotted line is strata of pumice and volcanic ash that piled up on the sea floor. Both strata are considered to have been created by a series of submarine volcanic eruptions. Please pay attention to the arrow-marked large rock slightly above the dotted line. As you can see from the stature of the people standing on the side of the big rock, the rock is over two meters in diameter. The big rock is a chilled bomb that I explained in Section 13 and has unique cracks created by thermal distortion due to its chilling when the rocks were dropped into the sea from the crater.

Now, I will think about how large volcanic bombs like this came to this place from the crater. The chilled bomb is surrounded by strata of fine pumice and volcanic ash. Even if such rock fragments and bombs run through the seabed in groups, it is physically impossible for them to carry volcanic bombs over two meters in diameter to a distant place. This is because volcanic bombs are so heavy that they sink to the bottom. Therefore, it is logical to think that these volcanic bombs jumped from the crater to the sea and directly dropped onto this place where pumice and volcanic ash were piling up.

I will explain an old story guessed from the strata on this cliff, as well as what I already mentioned in the previous section. Rock blocks and rock fragments that gushed out by submarine volcanic eruptions flooded down the slope in a group and formed the strata of subaqueous volcaniclastic flows below the dotted line in the previous photo. When this occurred, the inside of rock blocks remained hot. When the rock blocks reached this place, they were still 450 to 500 degrees Celsius. Subsequently, when pumice and volcanic ash spouted from the crater drifted in the sea, piled up in this place and began to form the strata above the dotted line, the arrow-marked large volcanic bomb dropped onto the place. Afterward, pumice and volcanic ash continued to pile up and cross laminations were created in the upper part of a pile of them under the influence of sea currents and waves. They currently form beautiful ripple-shaped patterns at the very top of the cliff and attract many tourists today.

The period of submarine volcanoes

A cliff around Tensodo on the Dogashima Coast. Beautiful strata created by submarine volcanoes can be seen.

The close-up view of the chilled bomb on the upper part of the cliff (see the photo below in the previous section)

16. The Shirahama Group (5): White cliff

There are many white cliffs formed by the strata of pumice spouted by submarine volcanic eruptions not only on the Dogashima Coast but also everywhere on the Izu Peninsula. Major examples include Shuzenji Spa, Shimoshiraiwa and Hiekawa in the city of Izu, Muroiwado in Matsuzaki Town, some places in Minami-Izu Town and the city of Shimoda and Enoura in the city of Numazu. The white cliffs except for the Dogashima Coast are primarily located in mountainous areas and do not attract much attention.

As a fundamental question, what is pumice? The real essence of pumice is volcanic glass containing many bubbles. Volcanic glass is a glassy rock that is formed when magma is chilled around the crater during a volcanic eruption. Bubbles are foams created after volatile elements contained in magma evaporated.

When a submarine volcano erupted, volcanic bombs, volcanic lapilli and volcanic ash, as well as pumice, must have been spouted from the crater. However, volcanic bombs are so large and heavy that they drop and sink first while being carried to a faraway place from the crater and are separated from other volcanic products. Meanwhile, volcanic ash continues to drift through the sea and flows away to a distant place in sea currents. Pumice is so light with bubbles in the beginning that it floats on the surface of the sea and drifts in the water. However, seawater gradually soaks into the inside of pumice, which causes pumice to end up sinking into the sea floor along with volcanic lapilli.

The strata on the white cliff seem to contain a pile of pumice in particular, but a close observation of it shows that the strata contain fine rock fragments. In fact, strata that piled up on the seabed are characterized by a combination of large pumice and tiny rock fragments. This is because the terminal velocity at which large pumice sinks in the water is almost equal to that of tiny rock fragments. Small pumice drifts away to a more distant place. In contrast, large rock fragments quickly sink into the ocean floor near the crater. Consequently, large pumice and tiny rock fragments coexist at a certain spot on the seabed. For the proportion of the size of pumice and rock fragments, they are large underwater and small on land. It is sometimes possible to judge if particular strata were created on the sea floor or on land by noting this proportion.

The period of submarine volcanoes

White cliffs like this can be seen everywhere in Izu. Many of them are strata formed by a thick pile of white pumice. The picture was taken around Hiekawa in the city of Izu.

Strata of pumiceous tuff distributed around Hiekawa in the city of Izu. The strata were formed after pumice gushed out by submarine volcanic eruptions had piled up on the sea floor. A close look at it shows that not only large pumice but also tiny rock fragments are contained.

17. The Shirahama Group (6): Volcanic neck

The Izu Peninsula has a mountainous terrain. There are several strange-looking mountains and rocks that stand out and attract particular attention. Those examples include Mt. Joyama and Mt. Katsuragiyama in the city of Izunokuni, Mt. Shimodafuji and Mt. Nesugatayama in the city of Shimoda and Mt. Eboshiyama and the Senganmon Rock in the Kumomi area, Matsuzaki Town (see Frontispiece 13 [above]). Many of these unique mountains are shaped like a bell with its sides being precipitous, but some have geologically unclear shapes.

These unique mountains and rocks were formed after magma, cooled and hardened just below a volcano, was washed by subsequent erosion and they are called volcanic neck. It is, so to speak, the "root" of a volcano. Its real substance is mostly a firm body of volcanic rock and many have contracted cracks due to cooling, such as columnar joints.

Volcanic necks are ubiquitous in volcanic areas all over the world, as well as Izu and other parts of Japan, just like other types of geomorphology of volcano. For example, Devils Tower (Wyoming, U.S.), a unique rock located at a spot on which a huge UFO landed at the climax of the movie *Close Encounters of the Third Kind*, is also another example of volcanic neck. Great columnar joints can be observed on the cliff on the sides of Devils Tower.

A volcanic neck is a kind of intrusive rock that was formed after magma got into strata, cooled and hardened. Intrusive rocks that have unclear geomorphology are distributed everywhere on the Izu Peninsula. However, large volcanic necks are almost limited to the Shirahama Group. Large-scale intrusive rocks are often quarried because a large amount of homogeneous rocks can be obtained from those rocks. Those examples include quarries at Okubo-no-hana in the city of Numazu and Mt. Shiratoriyama in the city of Izunokuni.

A platy intrusive rock that is vertical or sharply inclined is called dike. A dike can be easily found because it runs across the stripes of strata. Conspicuous dikes can be seen on the coastal cliffs around the west coast of Minami-Izu Town, Futo in Nishi-Izu Town and Ajiro in the city of Atami. There are a few platy intrusive rocks that are almost horizontal to strata and such intrusive rock is called sill. A beautiful example of sill can be seen on the coastal cliff at Tsumekizaki in the city of Shimoda.

The period of submarine volcanoes 45

Mt. Joyama, which rises along the Kanogawa River in the city of Izunokuni, is a major volcanic neck on the Izu Peninsula. The picture was taken by Yusuke Suzuki.

An example of dike running across the strata in the Koura area, Minami-Izu Town. The picture was taken by Yusuke Suzuki. Fontispiece 13 [below] is a close-up view of the lowermost part of this dike.

18. The Shirahama Group (7): First land

Most of the Shirahama Group is strata formed by submarine volcanoes, but there are a few exceptions. Those are strata formed by terrestrial volcanic eruptions.

Unlike on the sea floor, the air exists on land and some hot volcanic products fresh from the crater are exposed to the air. Then, oxygen in the air combines with iron within rocks under the influence of heat, which produces reddish brown iron oxide mineral called hematite. Where this mineral is produced in mass, the rocks turn into reddish. The rocks are literally "burned" and turn into red. Because of this, lava flows that ran down land and volcanic bombs and volcanic lapilli that dropped onto the ground turn into red in their parts that were exposed to the air for a long time.

Of course, redness is not the only characteristic of terrestrial volcanic products. Unlike eruptions on the ocean floor, lava flows created by terrestrial volcanic eruption are not broken by the hyaloclastic effects and many of them become plate-shaped slabs of rock. There are no chilled bombs and stratified stripes formed by sea currents and waves. As a matter of course, there are no fossils of creatures that inhabited the sea. Instead, strata containing paleosol and aeolian dust that pile up during the stoppage of eruptions are created. In addition, as noted in Section 16, there are many apparent differences, such as the relatively small proportion of pumice and rock fragments that fall down onto a place equally distant from the crater compared with the seabed.

These unique characteristics of terrestrial volcano can be widely seen in strata (the Atami Group) formed after the whole of Izu was uplifted and became land approximately one million years ago. However, as a result of a careful investigation into the Shirahama Group, strata with these characteristics of terrestrial volcano were discovered in several places, including Ajiro in the city of Atami and Umegi in the city of Izu. Such strata are spotted in a small range of a particular area.

This makes me imagine that at that time Izu was in an environment similar to that of the current geological environment around the Izu Shichito Islands. That is, it is conceivable that almost the whole of Izu was on the shallow sea floor and that volcanic islands were scattered around. It was the first formation of land in Izu.

The period of submarine volcanoes

An example of agglutinate that can be considered a volcanic product from terrestrial volcanoes. Hot volcanic bombs remaining in a state of melting dropped around the crater and piled up, collapsed as if soft rice cakes had dropped onto the ground. The whole of the stratum is reddish. The picture was taken around Ajiro in the city of Atami.

An example of stratum that can be considered a volcanic product from terrestrial volcanoes. The picture was taken in the Umegi area, the city of Izu.

Chapter 2

The path to a peninsula

19. Drilling into the sea floor around Izu (1): Deep-sea drillship

To closely analyze the history of earth in a particular area, it is often not enough to just conduct a geologic survey on the area. It is necessary to gain a deeper understanding by investigating into the geologic characteristics of neighboring areas and comparing the results. It was in 1989 that a fortunate perfect opportunity for such survey happened to come to me. At that time, I had just finished a post-doctoral geologic survey of Izu. I was invited to take part in an investigation into the marine geology around the Izu Shichito Islands as an onboard scientist of the deep-sea drillship *JOIDES Resolution.*

A deep-sea drillship is the vessel that is capable of drilling a hole into the sea floor, pulling up all drilled strata and rocks and conducting every investigation and analysis. The drillship *JOIDES Resolution* (with a displacement of 18,000 tons) drilled 1,797 holes (a total of 320,000 meters) into the seabed all around the world from 1985 to 2003 and produced great research results.

In the center of the ocean drilling research vessel is a derrick 61 meters in length. The crew puts down steel pipes (with a length of 28 meters) one by one from the derrick to the sea floor and drills a hole into the bottom of the sea by turning the pipes around by a powerful motor. The total extension of those pipes is long enough to drill through the seabed at a depth of thousands of meters. Of course, a pipe has an installed bit to drill into hard rocks on its tip. The mechanism of drilling is the same as that of terrestrial hot spring boring, but the scale is quite different.

It takes a maximum of dozens of days to finish drilling one single hole and if the ship drifts away during that period of time under the influence of waves and sea currents, the pipes break. That is why a total of 12 special screw propellers called thruster are installed around the ship and this special device enables the ship to maintain the same position by computerized control even in troubled waters.

The JOIDES Resolution is administered in accordance with the International Phase of Ocean Drilling (IPOD). Twenty onboard scientists or so were permitted to participate in one drilling project and the competition was tough. I had long yearned to participate in the project and finally, my long-cherished wish came true in the Leg 126 in 1989. Fortunately, this expedition was intended to investigate into the origin and history of a broad range of the geological basement around the Izu Shichito Islands.

The path to a peninsula

The dignified form of the deep-sea drillship *JOIDES Resolution*.

A group photo taken on the deck of the deep-sea drillship *JOIDES Resolution* together with other onboard Japanese scientists. At the back of the picture, you can see a drilling derrick in the center of the ship.

20. Drilling into the sea floor around Izu (2): International joint study

Our investigation around the Izu Shichito Islands aboard the deep-sea drillship *JOIDES Resolution* in the Leg 126 lasted for two months from mid-April to June in 1989. During that period of time, we neither disembarked nor stopped at a port. The voyage was just like a training camp with a total of 26 onboard scientists consisting of five Japanese researchers, including me, and other international researchers with many different nationalities and professional specialties. Of course, we spent the entire project in total English immersion in every situation ranging from onboard life to reporting at each investigation point.

During the project, the *JOIDES Resolution* operated around the clock without a break and all the crew, including onboard scientists, worked on twelve-hour shifts. Specifically, each onboard scientist analyzed strata and rocks pulled up from the ocean floor one after another according to his own assigned role and passed the results on to his partner with the same assigned duty every twelve hours. I played a role in measuring weak magnetism of the rocks as a paleomagnetist and my partner was an American researcher. Other roles included a sedimentologist who studied the organization and structure of strata, a paleontologist who studied fossils contained in strata, a petrologist and geochemist who conducted a chemical analysis of strata and rocks and a physical property scientist who measured the physical quality of strata and rocks. As I explained in Section 3, rock magnetism measurement provides significant clues to know the geologic age of strata and we were shouldered with heavy responsibility as rock magnetism researchers.

We faced many problems and difficulties during the two-month voyage. However, we completed our investigation into 19 drilling points (the total thickness of drilled strata was 2,122 meters) in the seafloor ranging from Aogashima to Tori-shima and obtained significant results for considering how the earth of Izu and its neighboring areas was formed.

In this connection, the *JOIDES Resolution* was a second-generation vessel used for IPOD. The third-generation deep-sea drillship *Chikyu* (with a displacement of 57,000 tons) was built by Japan and entered into service for an investigative expedition in September 2007. The *Chikyu* appeared and played a major role in the movie *Japan Sinks* (the remake version of 2006). I was much delighted to happen to catch a glimpse of the *JOIDES Resolution* in the same movie.

The path to a peninsula

A scene from the laboratory of the deep-sea drillship *JOIDES Resolution*. A massive cryogenic magnetometer using superconductivity and the author who was assigned to operate this device.

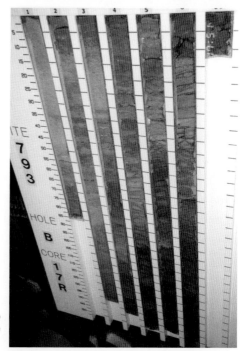

Samples drilled by the deep-sea drillship *JOIDES Resolution* in the seafloor near the Izu Islands.

21. Drilling into the sea floor around Izu (3): The northward drift of plates

In Section 1, I wrote that the earth of Izu travelled a long way from the south by plate motion and explained that the reason for this was based on a backward calculation from the current plate motion. In addition, in Section 10, I argued that the discovery of subtropical fossils from within strata in Izu was evidence showing that Izu used to be located in the South Seas. In fact, however, more direct evidence demonstrating where Izu used to be has been obtained. This result was gained by the method to measure weak magnetism of rocks, as mentioned in Section 3.

A geomagnetic field has almost the same form as the magnetic field of a bar magnet. Therefore, the direction of magnetic force is horizontal around the Equator and vertical around the Poles. The inclination of magnetic direction between the Poles and the Equator differs by latitude and the closer it is to the Poles, the sharper the angle is. The inclination around Japan is at latitude approximately 50 degrees. The reason why the magnetic needle of a compass appears to be horizontal is that it is forcibly made back to horizontal by placing a weight at the S-pole.

The direction and intensity of a geomagnetic field in old times when particular rocks were formed were recorded within those rocks as weak magnetism. Therefore, measuring weak magnetism on the rocks enables you to estimate the paleolatitude of a particular place where the rocks were formed on the basis of information about the inclination.

The rocks contained in strata distributed in Izu, the rocks drilled from the seabed around the Izu Shichito Islands by the deep-sea drillship *JOIDES Resolution* that I explained in the previous section and the rocks collected in other areas sitting on the Philippine Sea Plate all show an increase in latitude with time. This means that the whole Philippine Sea Plate that includes the Izu Shichito Islands, the Ogasawara Islands, the Mariana Islands and the broad stretch of seafloor around those islands, as well as Izu, continued to drift northward for more than 40 million years by plate motion. However, the latitude data obtained from strata in Izu lacked reliability due to the influence from rock alteration. It is probably because of this that Izu appears to have passed the Izu Shichito Ridge and to have drifted northward on graphs. My recognition is that the data obtained by the *JOIDES Resolution* were finally able to provide supplementary results worth relying on.

The path to a peninsula

The lines of magnetic force of a geomagnetic field. The closer to the North Pole from the Equator it is, the larger the inclination becomes.

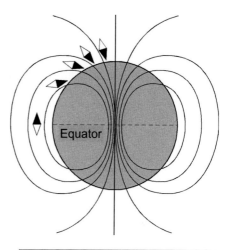

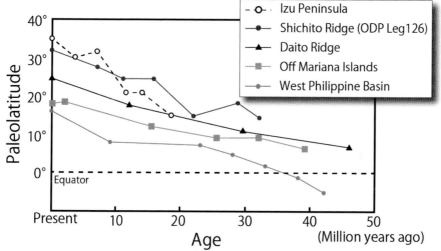

▲ Latitudes estimated from the magnetism of rocks collected on the Philippine Sea Plate. Only major results are shown on the graph. The horizontal axis means the geologic age of rocks.

◀ The logo mark of the deep-sea drillship *JOIDES Resolution*'s Leg 126 that I participated in. All the logos are chosen by voting for each and every voyage.

22. Izu's collision with the Japanese main island (1): Closed channel: The Ashigara Group

In Section 1, I wrote as follows: "The whole Izu Peninsula used to be a volcanic island (partly a submarine volcano) floating in the South Seas. It was 600,000 years ago that Izu collided with the Japanese main island and got peninsula-shaped." That is, in the times older than 600,000 years ago, what is Izu today used to be located far south of the Japanese main island and did not have its current peninsula shape. There was a paleochannel connecting the Suruga Bay with the Sagami Bay between Izu and the main island of Japan. The channel literally "closed" due to the accession and collision between Izu and the Japanese main island 600,000 years ago.

Where there is a sea, rock debris flows into the sea floor from its neighboring land and strata are formed. If the sea closed due to collision by plate motion, the strata must have been uplifted, become land and must still exist somewhere. Are there such conceivable strata anywhere around the current Izu Peninsula?

Hakone Volcano is situated at the joint of the Izu Peninsula. Hakone Volcano is an active volcano that began to erupt 600,000 years ago and is still continuing its active geothermal activity. The Sakawagawa River runs surrounding the northern side of the volcano. Major transportation routes, such as the JR Gotemba Line, Route 246 and the Tomei Expressway, run along the river. Starting from the foot of Mt. Fuji around Gotemba, the Sakawagawa River runs through the deep ravines in the Ashigara Mountains on the northern side of Hakone Volcano to the Ashigara Plain and flows into the Sagami Bay at the east of Odawara.

It was known for a long time that strata containing piles of gravel, sand and mud (the Ashigara Group) were widely distributed in the Ashigara Mountains. The strata are so thick that its total thickness combined even amounts to over 5,000 meters. In addition, the strata that were initially almost horizontal underwent heavy crustal deformation. In the beginning, no one knew why such strange strata were distributed there.

However, if you assume that the Ashigara Group is strata that piled up in the channel that used to be situated between the Japanese main island and Izu and that the group underwent heavy crustal deformation when the channel closed due to the collision between Izu and the Japanese main island, it can be a logical interpretation. It was around the early 1980s that researchers began to conduct various surveys and studies to prove that.

The path to a peninsula

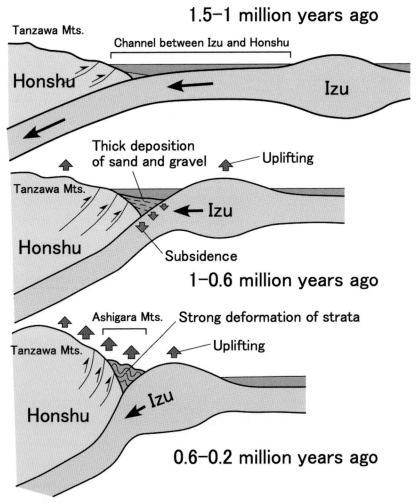

The process of collision between Izu and the Japanese main island (refer to the illustration in Section 2 as well).

23. Izu's collision with the Japanese main island (2): Closed channel: Landfill and uplift

Initially, the Ashigara Group was considered to be at least millions of years old from its well-solidified rocks. If this assumed geologic age is correct, it was even earlier than approximately one million years ago when Izu and the Japanese main island began to collide. In accordance with this line of reasoning, it is difficult to think that the Ashigara Group was the strata that caused the channel to close immediately before the collision. It was in the mid-1980s that this issue of discussion was resolved. During that period of time, there was rapid progress in a great variety of studies to clarify the process of collision between Izu and the Japanese main island, including my doctoral dissertation.

First of all, the types of planktonic fossils contained in the strata and paleomagnetic measurement of rocks revealed that the Ashigara Group was approximately two million to 700,000 years old, much younger than had been expected. This confirmed that the Ashigara Group was exactly the strata that had been formed when Izu and the Japanese main island collided.

The next notable point is the estimated sea depth of the place where the Ashigara Group piled up. Oceanic microorganisms can be divided into the group that drifts in the sea (plankton) and the group that lives on the ocean floor (benthos). The latter differs in species by the depth of the seabed. Therefore, it is possible to estimate paleobathimetry by examining the species of fossils contained in strata.

By this method, the paleobathimetry of the place where the Ashigara Group had been formed could be estimated to have been 2,000 to 1,000 meters approximately two million years ago and 400 to 100 meters approximately one million years ago. That is, it emerged that the Ashigara area used to be the deep sea two million years ago, but was subsequently rapidly filled in by rock debris and continued to uplift after it had become land. Currently, the Ashigara Mountains are 1,000 meters above sea level, which means that the level of uplift from two million years ago amounts to 3,000 meters.

This rapid landfill from the sea and uplift can be logically explained if you assume that the channel between Izu and the Japanese main island became land by landfilling, was subsequently constricted by the collision between Izu and the Japanese main island and became mountains.

The path to a peninsula

A thick conglomerate formed by gravel of the Ashigara Group. The strata that must have been initially horizontal are currently erecting almost vertically due to subsequent crustal movements. The picture was taken in a mountain around the Surugaoyama Station on the JR Gotemba Line.

The Kannawa Fault in Oyama Town, Shizuoka prefecture, which is one of the places where Izu and the Japanese main island collided. The strata on the side of the Japanese main island (Honshu) and the strata on the side of Izu meet across from the fault.

24. Izu's collision with the Japanese main island (3): The last sea

Around one million years ago when Izu and the Japanese main island began to collide by plate motion and the channel between them was being filled in by the Ashigara Group, a rare incident was occurring in Izu as well. It was a phenomenon in which strata formed by normal mud, sand and gravels piled up.

You should not think that such strata are commonplace. Most strata on the Izu Peninsula are terrestrial and submarine volcanic products and it is very difficult to find non-volcanic deposits. This is because Izu used to be a submarine volcano (partly a small volcanic island) far away from the Japanese main island for a long time. Non-volcanic deposits are rock particles created after land was cut by rainwater and weathering. Those particles flow into the rivers and pile up in the nearby sea, which leads to the formation of strata. That is, the presence of such strata containing mud, sand and gravels means that there used to be a large land area nearby.

Strata formed by piles of these non-volcanic deposits are distributed in limited places in the mid-northern area of the Izu Peninsula, specifically, in the mountains in Ono, Jo, Umegi and Ikadaba in the city of Izu (see Frontispiece 16). An examination of the fossils of oceanic microorganisms contained in muddy strata in Jo and Umegi (the Yokoyama Siltstone) revealed that they were approximately 1.2 million years old and were species that lived on the seabed with a depth of 200 to 600 meters. Gravel strata named Ono Conglomerate are piled up on the Yokoyama Siltstone. The Ono Conglomerate can be considered strata of submarine fan formed near the mouth of a river. That is, this shows that shortly after a bay or a trough had been born in some areas of Izu, they were filled in by strata of mud and gravels. This process is similar to the history (rapid landfill from the deep sea and transformation into land) of the Ashigara area during the same period of time and can also be associated with the collision between Izu and the Japanese main island.

After these non-volcanic deposits had intensively piled up in particular areas, strata originating from the sea disappeared from Izu and there were nothing but terrestrial volcanic products. That is, the "last sea" in Izu disappeared around one million years ago and the transformation of the whole of Izu into land was completed.

The path to a peninsula

Strata of the Ikadaba Sandstone. This sand piled up in the sea that used to be located here around one million years ago. The picture was taken around the Ikadaba area in the city of Izu.

Strata of the Yokoyama Siltstone. The picture was taken around the Hatsuma area in the city of Izu.

25. Izu's collision with the Japanese main island (4): The indentation of Izu

After the collision between Izu and the Japanese main island, the channel between them underwent rapid landfill and uplift and strata were folded by heavy crustal deformation. The traces of these vertical deformations are likely to be left in the characteristics and structures of strata and they can easily be examined by an ordinary geologic method. However, in the case of massive horizontal movements and the tectonic rotation of a whole area, it is difficult to even know their occurrence unless you use a special method. One example of it is a large-scale horizontal movement by latitude changes (Izu's northward drift by plate motion), as explained in Section 21.

In the meantime, it is fully likely that Izu and its neighboring areas underwent a large-scale tectonic rotation along with the collision between Izu and the Japanese main island. An effective means to detect evidence for this is the method to measure weak magnetism on rocks, which I repeatedly explained in this book.

The angle that shows how much the direction of a geomagnetic field is vertically inclined is called *inclination*. By the same token, the angle that shows how much the direction of a geomagnetic field is horizontally deviated from the geographic north is called *declination*. The direction and intensity of a geomagnetic field in old times when strata and rocks were formed were recorded within those strata and rocks as weak magnetism. Therefore, measuring weak magnetism on the strata and rocks enables you to estimate how much a particular place where the strata and rocks were formed rotated afterward on the basis of information about the declination.

An examination of Izu and its neighboring areas by this method revealed that a major tectonic rotation had not occurred in Izu at least after five million years ago with the exception of a few areas. However, surprising tectonic rotations occurred in areas surrounding the Izu Peninsula. A clockwise rotation was detected with the northeastern side of Izu (Oiso Hill, Tanzawa Mountains and Miura Peninsula) and a counterclockwise rotation was detected with the northwestern side of Izu (Kambara Hill). A major tectonic rotation with an angle of more than 50 degrees was even detected with some places. In addition, it was discovered that those tectonic rotations occurred immediately after the collision between Izu and the Japanese main island. This discovery revealed that after the collision with the Japanese main island, Izu indented the main island, which pushed and rotated its neighboring areas.

The path to a peninsula

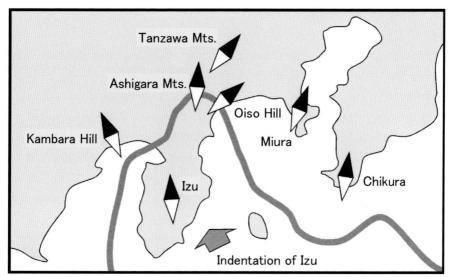

This illustrates the average values of the declinations of rock magnetism in several areas by the bearings of magnetic needles. How much each area rotated is shown by the angle of deviations from due north. The arrow mark means the direction of movements of Izu relatively to the Japanese main island and the bold gray line means the position of a plate boundary.

A bird's-eye view of the Oiso Hill. The basement of this hill rotated clockwise by an angle of 50 degrees due to the indentation caused by the collision between Izu and the Japanese main island.

26. Diving in the bottom of the Suruga Bay (1): *SHINKAI 2000*

The Suruga Bay stretches on the western side of the Izu Peninsula. Both sides across from this bay greatly differ in geologic characteristic. Almost all strata distributed in Izu are made up of volcanic products with some exceptions. In contrast, many of the strata distributed on the west coast of the Suruga Bay are formed by piles of non-volcanic deposits. The plate boundary between the Philippine Sea Plate on which Izu sits and the plate of the Japanese main island, part of which the west coast of the Suruga Bay is, must pass across somewhere around Mt. Fuji on land. However, it is difficult to directly observe this plate boundary because it is buried deep beneath the lava flows and debris flows from Fuji Volcano.

Now, I will take a look at the sea floor. For the marine geomorphology of the Suruga Bay, the protuberance of the seabed on the Izu Peninsula side and the protuberance of the seabed on the west coast of the Suruga Bay side appear to meet along a submarine gorge off the west coast of Matsuzaki Town. This spot is 1,850 meters deep. It was expected that diving in this spot would make it possible to directly observe the plate boundary and that an examination of the characteristics and geologic age of strata distributed there might reveal some historical process of the Suruga Bay and Izu. It was in 1991 that an underwater survey of the submarine gorge was carried out in expectation of this possibility. *SHINKAI 2000*, a submersible Japan boasts of, was used for this marine investigation project.

SHINKAI 2000 is a manned research submersible (a length of 9.3 meters, a weight of 24 tons and a carrying capacity of 3 people) constructed in 1981 by the Japan Marine Science and Technology Center (currently, Japan Agency for Marine-Earth Science and Technology, JAMSTEC). As its name shows, this submersible can dive to depths of 2,000 meters. Currently, it has virtually retired from service and has been replaced by *SHINKAI 6500*, which can dive to depths of 6,500 meters. In this connection, in the movie Japan Sinks (the remake version of 2006), the real things of these two submersibles appear with the changed names of Wadatsumi 2000 and Wadatsumi 6500, respectively, and show a great performance.

On the morning of October 29, 1991, *SHINKAI 2000* was lowered from its mother ship Natsushima into the Suruga Bay waters 20 kilometers off the coast of Matsuzaki, which marked the beginning of the Dive 579. It was boarded by two pilots and one onboard scientist (the author).

The path to a peninsula

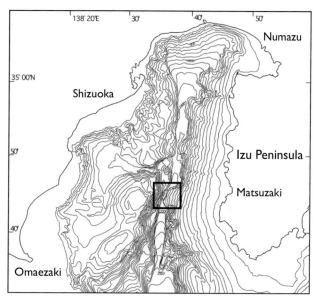

Marine geomorphology of the Suruga Bay. The interval of the isobaths is 100 meters. The square means the area in which the *SHINKAI 2000* Dive 579 was carried out. Bathymetrical data are after the Japan Coast Guard.

Operators are preparing *SHINKAI 2000* for the Dive 579 on the deck of the mother ship Natsushima off the coast of Matsuzaki in Izu.

27. Diving in the bottom of the Suruga Bay (2):
The Dive 579 launched

On the morning of October 29, 1991, the manned research submersible SHINKAI 2000, which I was aboard, began the Dive 579 on the sea floor of the Suruga Bay 20 kilometers off the coast of Matsuzaki in Izu. The purpose of this marine survey was to directly observe the plate boundary between the Philippine Sea Plate on which Izu sits and the plate on the side of the Japanese main island and to investigate into the historical process that the Suruga Bay and Izu went through.

The rules prohibited night-time dives and all operations ranging from the preparation of the submersible to the housing in its mother ship were conducted in the daytime. We spent six hours or so in the submersible and less than four hours for a survey of the seafloor in the Suruga Bay. The submersible was battery-powered and was not equipped with a heating device for saving electricity. That is why the temperature of the submersible in the deep sea was five degrees or so and we were prepared cold-protection clothes. Of course, the vessel had no toilet installed and the cold-protection clothes were set with a portable toilet for emergency.

Just a few minutes after the dive had been launched, I found myself surrounded by the darkness. I was completely in a world of darkness in which I could see only marine snow shining in the light. I saw deep-sea fish and jellyfish swimming beside the round window of the submersible from time to time. I felt as if I had gone out for a geologic survey with only a flashlight in my hand at midnight. In addition, even if I caught sight of a stratum I could not directly touch it by hand. If I wanted to collect rocks, I had no other choice but to ask the pilots to take it by magic hand. This operation looked very difficult and it took even several minutes to pick up just one rock.

About one and a half hours after the dive had begun, we reached the bottom of the Suruga Bay. The tidal current from the south was 0.9 knot (approximately 50 centimeters per second) and we reached the sea floor 1,850 meters deep having much difficulty well piloting the vessel. I saw ripple marks (ripples formed on the surface of sand) showing a sign of strong tidal currents on the surface of the sea floor. The presence of strong tidal currents in such deep waters was reminiscent of the opening scene of the movie *Japan Sinks* (1973). In this movie, the submersible *Wadatsumi* is facing the danger of being wrecked under attack from turbidity currents, which I explained in Section 8. However, fortunately, the submersible does not encounter

The path to a peninsula 67

strong tidal currents in places other than the opening scene and *SHINKAI* was able to complete its survey mission in safety.

In this connection, the sea floor of the Suruga Bay was unexpectedly dirty, although it was deep waters. A supermarket shopping bag was drifting just beside crabs walking on the sea floor.

A cliff formed by old submarine volcanic products around the waters 1,800 meters deep in the Suruga Bay. At the front is a deep-sea fish that appears to be a kind of grenadier.

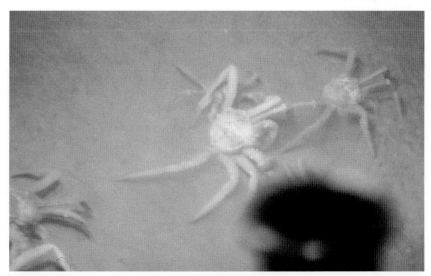

Crabs inhabiting the seabed of the Suruga Bay. The shadow at the front is a part of the submersible.

28. Diving in the bottom of the Suruga Bay (3): Sinking sea floor

In the Dive 579 by the submersible *SHINKAI 2000*, carried out in October 1991 in the Suruga Bay off the coast of Matsuzaki, after the vessel had first reached the deepest spot 1,850 meters deep, it moved along a route to continuously observe strata distributed to depths of 1,600 meters, gradually going up the slopes on the side of Izu. For significant results obtained from this marine survey, we confirmed that there existed old submarine volcanic products and shallow marine sandstone covering those volcanic products.

The submarine volcanic products were strata formed by piles of chilled bombs and lava flows and had similar characteristics to the Shirahama Group, which I explained in this book. However, strata containing the same sand and mud as the ones widely distributed on the west coast of the Suruga Bay were observed on the slope on the side of Shizuoka just 200 meters away from the spot where the submarine volcanic products were seen. That is, it was demonstrated that the strata peculiar to Izu and the strata peculiar to the Japanese main island are distributed close to each other at a slight distance and that the plate boundary between the Philippine Sea Plate on which Izu sits and the plate on the side of the Japanese main island pass across the deepest spot in the Suruga Bay, as was expected.

Meanwhile, the shallow marine sandstone covering the submarine volcanic products is calcareous sandstone containing many fossils of oceanic microorganisms that live in the waters 30 to 100 meters deep and can be considered to be approximately one million years old. This sandstone is currently distributed at depths of 1,750 meters, which means that it sank by as much as approximately 1,700 meters for one million years. That is, part of the strata that form Izu at this spot are about to subduct under the Japanese main island by plate motion.

The Philippine Sea Plate on which Izu sits is still continuing to drift northwestward at a few centimeters per year and the Suruga Bay is getting smaller. A massive amount of rock debris are flowing into the Suruga Bay from many rivers, including the Fujikawa River, the Abekawa River and the Oigawa River. As I explained in Sections 22 and 23, the channel that used to be located around the Ashigara Mountains on the northern side of Izu was rapidly filled in by rock debris and closed due to the accession and collision between Izu and the Japanese main island under the influence of plate motion. The same phenomenon is now occurring in the Suruga

The path to a peninsula

Bay. The bay will have completely closed in hundreds of thousands of years. Then, definitely, it will no longer be possible to call Izu a *peninsula*.

Old submarine volcanic products observed around at depths of 1,800 meters in the Suruga Bay.

Shallow marine sandstone observed around at depths of 1,600 meters in the Suruga Bay. Fossils of shell can be seen within some of them.

Chapter 3

The period of terrestrial large volcanoes

29. A long range of multiple volcanoes: Polygenetic volcano

As noted in Section 27, marine strata disappeared from Izu approximately one million years ago. It means that the whole of Izu became land. In this situation, large volcanoes came into being and their geologic formations as mountains remain as they are even today. There are thirteen major large volcanoes and by size, they are Amagi Volcano, Taga Volcano, Daruma Volcano, Tanaba Volcano, Usami Volcano, Yugawara Volcano, Nekko Volcano, Tenshi Volcano, Ita Volcano, Jaishi Volcano, Chokuro Volcano, Osezaki Volcano and Nanzaki Volcano.

Most of these volcanoes are called polygenetic volcanoes. Polygenetic volcanoes repeatedly erupt from almost the same spot for tens of thousands of years to hundreds of thousands of years with some intervals of stoppage and results in getting large forms. Fuji Volcano, Hakone Volcano, Ashitaka Volcano and Izu Oshima Volcano around Izu are also polygenetic volcanoes.

Fuji Volcano, Hakone Volcano and Izu Oshima Volcano are categorized as active volcanoes and these three volcanoes may erupt in the future. However, somehow, the above-mentioned thirteen polygenetic volcanoes on the Izu Peninsula had stopped erupting by 200,000 years ago and none of them are categorized as active volcanoes.

On the Izu Peninsula after 150,000 years ago, large volcanoes that had "died down" were replaced by small volcanoes represented by Omuroyama Volcano and these small volcanoes began to erupt everywhere on the peninsula. Consequently, a group of small volcanoes (the Izu Tobu Volcano Group) were formed and this group has remained to this day. These small volcanoes stop erupting from the same crater after they erupt only once and form a whole new crater next time they erupt. This type of volcano is called monogenetic volcano in comparison with polygenetic volcano.

The magma of the Izu Tobu Volcano Group still continues its occasional underground activity and causes earthquake swarms off the east coast of the Izu Peninsula. In July 1989, the magma caused a submarine eruption off the coast of Ito and created a new monogenetic volcano, Teishi Knoll. Based on these facts, the Izu Tobu Volcano Group is counted among active volcanoes.

To sum it up, volcanic activities in Izu after it became land can largely be divided into two on the boundary of approximately 150,000 years ago: large (polygenetic) volcanic activities earlier than 150,000 years ago and small (monogenetic) volcanic activities later than 150,000 years ago.

The period of terrestrial large volcanoes

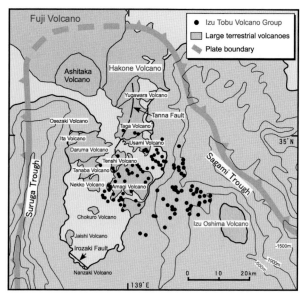

The distribution of volcanoes that erupted in Izu and its neighboring areas after approximately one million years ago. They are divided into large volcanoes and small volcanoes (the Izu Tobu Volcano Group). The Izu Tobu Volcano Group is also distributed on the sea floor between the Izu Peninsula and the Izu Oshima Island. The interval of the isobaths of the waters is 500 meters.

A bird's-eye distant view of Mt. Amagisan, the highest mountain on the Izu Peninsula, seen from its northeastern side. The pudding-shaped mountain at the left front and the lake at the right bottom are Omuroyama Volcano and Lake Ippekiko, respectively, both of which belong to the Izu Tobu Volcano Group.

30. A long range of multiple volcanoes: Lost summit

A close look at the line of mountains in Izu on a topographical map shows that the ridge line that can be said to be the backbone of the peninsula is shaped like the alphabet J. First, a north-south ridge line runs through Hakone Volcano, the Jukkokutoge Pass, the Kameishitoge Pass and the Hiekawatoge Pass to Amagi Volcano. This ridge line begins to bend westward in Amagi Volcano, then runs through the southernmost Amagitoge Pass and bends northward. The ridge line also runs through the Nishinatoge Pass and the Funabaratoge Pass, then forms a north-south line again and goes to Daruma Volcano.

This J-shaped ridge line also forms a primary watershed. The inside of the J-shaped ridge line is the Kanogawa River and the rainwater that poured down within this range eventually meet together in the Kanogawa River and runs into the Suruga Bay at around Numazu. On the other hand, the rainwater that poured down within the range outside of the J-shaped ridge line runs into the Suruga Bay and the Sagami Bay through many small rivers, such as the Nishinagawa River and the Kawazugawa River.

This J-shaped ridge line was exactly created by particularly large seven volcanoes (Yugawara Volcano, Taga Volcano, Usami Volcano, Amagi Volcano, Nekko Volcano, Tanaba Volcano and Daruma Volcano in order of tracing the J-shaped ridge line) of the thirteen large volcanoes in Izu, mentioned in the previous section. That is, because these seven volcanoes lined up in a J-shaped form, it inevitably created the same-shaped ridge line.

These seven large volcanoes must have had beautiful shapes close to a conic form when they repeatedly erupted, if not as beautiful as that of Fuji Volcano. However, hundreds of thousands of years passed after they had stopped erupting and lost most of their original forms due to erosion during that period of time. Their erosion on the side of the sea is particularly remarkable. For example, Taga and Usami Volcanoes have already lost their original half eastern sides, which sank into the Sagami Bay. For each of the seven volcanoes, the slope on the side of Nakaizu is gradual and retains clear volcanic landforms, such as lava flows. However, the slopes on the sides of the Suruga and Sagami Bays are steep and hardly retain volcanic landforms. Of course, their summits have also already been lost and it is even uncertain where the previous peaks were.

However, erosion is a perfect phenomenon to observe the internal structure of

The period of terrestrial large volcanoes

volcanoes. In particular, various volcanic products created by eruption can often be seen on the cliff along the coastline. In the following section and beyond, I will take a look at geologic formations, volcanic products and landforms created by eruption with a focus on these thirteen volcanoes.

A long range of mountains that form the backbone of the Izu Peninsula was created by volcanoes. A three-dimensional map was drawn by digital elevation data developed by the Geospatial Information Authority of Japan and the mapping software Kashmir 3D.

The Omigawa River (at the front) and the Kanogawa River (at the back) run quietly through the middle part of Izu. The mountain at the left back is Daruma Volcano and over it is the Suruga Bay.

31. Yugawara Volcano, Taga Volcano and Usami Volcano

Along the Sagami Bay on the northeastern side of the Izu Peninsula are lines of hot spring towns, such as Yugawara, Atami, Taga, Ajiro and Usami, from the north. The Yugashima Group and the Shirahama Group, which are old strata, as explained in the previous sections, are distributed in the lowland areas in these towns. However, it was known long ago that there were massive piles of lava flows and volcanic ash peculiar to terrestrial volcanoes in the mountains at the back of the towns.

In the 1930s, a young geologist and petrologist from Tokyo began to conduct a survey of this area. His name was Hisashi Kuno (1910-1969). He later became a professor at the University of Tokyo and President of the Volcanological Society of Japan and also became a globally distinguished prodigious petrologist that Japan had ever produced.

He organized and classified the above-mentioned volcanic products as the three major volcanoes, Yugawara Volcano, Taga Volcano and Usami Volcano, from the north. After his survey, I also carried out an investigation into Usami Volcano and discovered that it was an old volcano that had repeatedly erupted from approximately one million years ago to 500,000 years ago. Yugawara and Taga are slightly younger than Usami Volcano, probably 800,000 to 300,000 years old.

Because most of volcanic products from these three volcanoes are distributed to the west of hot spring towns each volcanic name shows, probably, some are doubtful about the volcanic naming. However, Kuno was insightful enough to see through the fact that the eastern half sides of each of the three volcanoes had been lost by erosion and employed the names of Yugawara, Taga and Usami as places close to the center of the volcanoes.

For the current landforms of the city of Izunokuni and the city of Izu, a gradual slope stretches from the ridge line of mountains along which the Izu Skyline runs to the lowland areas along the Kanogawa River. This is the remnant of geomorphology that formed mountain foots westward from the summits of the three volcanoes of Yugawara, Taga and Usami, which probably used to be well beyond 1,000 meters above sea level.

In addition to these three volcanoes, Kuno published a succession of unique research results about the origin and nature of the Tanna Fault that caused the Kita Izu Earthquake of 1930 and the cause of the Ito Earthquake Swarm of 1930. I will explain those things in later sections.

The period of terrestrial large volcanoes

Volcanic products from Taga Volcano observed on the cliff along the Izu Skyline. Lava flows form layers.

The city of Ito seen from the south. At the back of the city stretches a gradual range of mountains formed by Usami Volcano and Taga Volcano.

32. Amagi Volcano and Tenshi Volcano

Amagi Volcano, the highest mountain on the Izu Peninsula, is a major large volcano representing Izu and was formed by eruption 800,000 to 200,000 years ago. The slope on its southern side facing the Sagami Bay has already been deeply eroded and hardly retains its original landforms, but the tableland in the Midaka area, Kawazu Town, can be considered to be a flat land that corresponds to the surface of original lava flows.

In the meantime, on the sides of the cities of Izu and Ito that form the northern half side of Amagi Volcano stretches a slope that forms a gradual mountain foot peculiar to volcanoes. The slope retains many clear landforms of lava flow. Particularly spectacular are a T-shaped tableland on the southern side of the Hiekawa River from the Hiekawa area, the city of Izu, to the Hatsuma area in the same city and a long and narrow tableland between the Omigawa River and the Jizodo River east of the Himenoyu area in the city of Izu. These tablelands have landforms created after lava flows as thick as 50 meters ran northward.

Like other large volcanoes in Izu, Amagi Volcano lost considerable parts (especially the southern side) of its original form by erosion and the position of its original summit is unclear. The current highest point is Banzaburodake 1,406 meters above sea level, but it is conceivable that its original summit used to be somewhere further to the south and was nearly 2,000 meters above sea level. Hatchoike Pond, which is located to the east of the Amagitoge Pass, is commonly said to be the crater lake of Amagi Volcano, but it is a depression formed after the innermost part of the valley caved in due to the displacement of an active fault and is not the crater (see the photo below in Section 65).

For Amagi Volcano, you can observe many new volcanic landforms, such as crater, small volcano and lava flow. Those landforms used to be considered to be part of Amagi Volcano for some time, but currently, they are considered to be part of the Izu Tobu Volcano Group.

On the northern side of Amagi Volcano is a hilly terrain sandwiched between the Kanogawa River and the Omigawa River and the hilly area forms mountain blocks independent from Amagi Volcano. Volcanic products, such as lava flow with terrestrial volcanic characteristics, are distributed in this hilly area and are named Tenshi Volcano after Mt. Tenshi 608 meters above sea level, the highest point of the hilly area.

The period of terrestrial large volcanoes 79

Tenshi Volcano is an old volcano formed by eruption approximately one million to 400,000 years ago. Most of its original volcanic landforms have been lost due to erosion and barely retains its remnants of volcano on the flat surface of the Shuzenji Country Club.

A gradual slope stretching from Mt. Misujiyama (upper right) to Cape Inatorimisaki (the peninsula on the left) in Higashi-Izu Town. It is a landform that corresponds to the southern slope of Amagi Volcano.

Lava flows from Amagi Volcano seen along a woodland path on the northern side of the Amagitoge Pass. You can see beautiful platy joints formed by contraction in time of chilling

33. Daruma Volcano, Ita Volcano and Osezaki Volcano

Daruma Volcano, which rises spectacularly high in the northwestern part of the Izu Peninsula, is a major large volcano in Izu along with Amagi Volcano and was formed by eruption one million to 500,000 years ago. On the western slope facing the Suruga Bay is a ravine that was deeply cut out by erosion and at its exit is the port of Heda (see Frontispiece 8 [above]). The volcano's original summit has already been lost and the current summit (982 meters above sea level) along the Nishi-Izu Skyline is just the highest point of the existing mountain form. It is thinkable that its original grand form probably nearly 1,300 meters above sea level used to stretch its foot to even the Suruga Bay off the coast of the port of Heda to the west and off the coast of Nishiura to the north.

Meanwhile, on the eastern slope of Daruma Volcano stretches a gradual slope that forms its original volcanic landforms to the area around Shuzenji. Probably, anyone who has driven from Shuzenji to Heda finds that there is a series of gradual upward slopes with few curves from Shuzenji to the Hedatoge Pass. This is because this route corresponds to the foot of Daruma Volcano. However, once you pass the Hedatoge Pass, you will find that the road comes to a winding sharp downhill. This is because the path goes down to a steep ravine cut out by erosion.

Daruma Volcano borders Ita Volcano and Osezaki Volcano on its northwestern side. Ita Volcano, which erupted from 800,000 to 400,000 years ago, is a little younger than Daruma Volcano, but it hardly retains its original mountain form due to huge erosion. Like Daruma Volcano, a large ravine created by erosion stretches to the west coast and at its exit is the village of Ita.

Osezaki Volcano is a volcano whose volcanic products can be observed only in the mountains south of Cape Osezaki. Most of its mountain form was lost by erosion and it is uncertain how large the volcano originally was. Because Osezaki Volcano is covered with volcanic products from Ita Volcano, it can be considered to be a little older than Ita Volcano.

You can well observe piles of volcanic products ejected from the above-mentioned three large volcanoes on the coastal cliffs from Osezaki to Toi (see Frontispiece 8 [below]). In particular, those volcanic products can clearly be seen from ships, so if you board the Suruga Bay Ferry and the high-speed ship *White Marine*, I recommend that you take a close look at the patterns of strata on the cliff.

The period of terrestrial large volcanoes

Daruma Volcano seen from its western side. A deep ravine has been formed by erosion and at its exit is the port of Heda (the town at the front of the picture).

Hugely eroded Ita Volcano. The Ita community and the Myojinike Pond can be seen at the bottom of the ravine to the west of the mountain.

34. Tanaba Volcano, Nekko Volcano and Chokuro Volcano

If you go down the Nishi-Izu Skyline to the south from Daruma Volcano, you will come to the Funabaratoge Pass. The Funabaratoge Pass is an important transportation route connecting Nakaizu with Nishi-Izu and the traffic is heavy. The Nishi-Izu Skyline ends here, but the Shizuoka prefectural road Nishi-Amagi Highland Line, which runs southward along the ridge line, opened in 1999. This road runs through Mt. Tanabayama (753 meters above sea level) and ends at the Nishinatoge Pass, across which a prefectural road connecting Yugashima with Nishi-Izu Town runs.

The ridge line also continues to the south from the Nishinatoge Pass and comes to the Amagitoge Pass through Mt. Nekkodake (1,035 meters above sea level). However, there are no car roads here and hikers just sometimes visit the place to enjoy mountain-ridge traversing. In addition, there is also a ridge line that branches southward from halfway between Mt. Nekkodake and the Amagitoge Pass and the ridge line goes through Mt. Saruyama to Mt. Chokuro (996 meters above sea level). Mt. Chokuro is a mountain that you can see to the northeast from Matsuzaki Town.

The mountains forming these ridge lines were created by piles of volcanic products from old terrestrial large volcanoes just like Daruma Volcano and Amagi Volcano. These mountains are divided into Tanaba Volcano, Nekko Volcano and Chokuro Volcano from the north on the basis of survey results.

Like Daruma Volcano on its northern side, Tanaba Volcano is an old volcano most of whose western side has been lost by erosion. At the exit of a large valley created by erosion is Toi Town. On the other hand, a gradual slope that can be regarded as a remnant of an old volcano barely remains on the eastern side (Yugashima side). Tanaba Volcano was formed from two millions to 800,000 years ago.

Nekko Volcano, which is adjacent to Tanaba Volcano on its southern side, was created by eruption one million to 600,000 years ago. Nekko Volcano's western (Ugusu) and southwestern (Nishina) sides are steep by erosion. However, it retains a gentle slope peculiar to volcano on its northeastern side (Yugashima side) and the Amagi Ranch has been built by using that flat landform.

Chokuro Volcano is also an old volcano created by eruption around 600,000 years ago. Most of its mountain form has been lost by erosion and retains a gradual slope just around Mt. Chokuro. Mt. Saruyama (1,000 meters above sea level), which is located north of Mt. Chokuro, is also formed by lava flows and may have been part of Nekko Volcano or Chokuro Volcano.

The period of terrestrial large volcanoes

A bird's-eye view of Toi. The mountain at the back is Tanaba Volcano. A distant view of the range of mountains is Amagi Volcano.

The cross section of thick lava flows from Nekko Volcano. The picture was taken around the Nishinatoge Pass.

35. Jaishi Volcano and Nanzaki Volcano

If you go down to the south along Route 136 from Matsuzaki Town, you will be driving along the coast for a while. However, then you go up a mountain just past Kumomi and will be driving on a gently sloping highland until you come to around the Koura area in Minami-Izu Town. The road around this area is a scenic spot called the Margaret Line. If you change to the west halfway through the road and go down to the coast, you can visit Cape Hagachizaki, which is well-known for its wild monkeys. The volcano that created this highland is called Jaishi Volcano after the name of its surrounding area.

Jaishi Volcano is an old volcano formed by eruption from 1.4 million to 1.3 million years ago. It is deeply affected by erosion and it is unknown how large it originally was and where its summit was. The landform of this volcano forms a gradual hill peculiar to volcano and its highest point is at 520 meters.

Now, if you go down southward from Koura along Route 136, you will be heading toward Cape Irozaki from Sashida past Mera. If you pass Nakagi and change the direction to the east toward Cape Irozaki, overlooking a precipitous cliff along the coastline on your right, you will find that the road comes into a small highland 50 to 100 meters above sea level.

This highland is called Ikenohara and has an elegant landscape like a small terrace facing the coastal cliff in the middle of a steep landform around Cape Irozaki. This place is famous for its day lilies and is managed and maintained as the Oku Iro Day Lily Park. Although it is a small terrace less than one tenth of the highland formed by Jaishi Volcano, this landform is also a product of volcano. Judging from its size, I cannot help hesitating to count this volcano among the major terrestrial large volcanoes in Izu. Because its southern side has been eroded into a sea, its original size is unclear.

This volcano created by eruption approximately 400,000 years ago is called Nanzaki Volcano. The reason why this volcano is known among some experts is that its rocks are extraordinarily unique. Although its rocks look quite normal, they are categorized as Alkali Basalt from the perspective of chemical element and are volcanic rocks that typically appear far away from the subduction zone of plate. It is uncertain why such extraordinary rocks spouted in this place.

The period of terrestrial large volcanoes

A port in the Mera area, Minami-Izu Town. The tableland seen over the bay is a landform created by Jaishi Volcano.

Ikenohara to the west of Cape Irozaki. A small terrace landform created by Nanzaki Volcano.

36. Volcano that created glass

If you go down to the south along Route 136 from Toi, you will be passing a height along the coast just before you get to Ugusu. This area near famous Cape Koibitomisaki is Koshimoda in the city of Izu. The landform around this area is gradual and is, on the whole, gently inclined downward to the west. This landform is the remnant of erosion-free part of the foot of a volcano that used to be situated on the eastern side.

This volcano does not have its name, but volcanic products, such as lava flows remaining around Koshimoda, are named Koshimoda Andesites. This volcano was considered to be older than two million years ago for some time. However, judging from its characteristics of terrestrial eruption and fresh nature of rocks, the volcano should be regarded as being as old as Tanaba Volcano, which is adjacent on its eastern side, or a little older than Tanaba Volcano, that is, two millions to one million years old.

Around the original summit of this volcano was a large geothermal field. Volcanic rocks underwent geothermal alteration by hot spring water underground this geothermal field and many mineral deposits were created. Among those deposits is the Izu Silica Stone Mine, which is located in a mountain three kilometers east of Koshimoda. Silica stone is mostly made of quartz with nearly the same composition as amethyst. It is useful as a material for plate glass and building material and large-scale mining was carried out. The whole mountain is dyed silica-stone white due to open-air mining and could clearly be seen from around the city of Shizuoka across from the Suruga Bay on a sunny day. Currently, it is not so conspicuous due to afforestation, but it is probable that gas smokes were spouting in the old age when the volcano used to be active.

This deposit of silica stone made Ugusu very famous among experts. They regard Ugusu as a synonym for silica stone. The beautiful yellowish white cliff along Cape Koganezaki, which is also located in the Ugusu area, also underwent the same geothermal alteration that created the silica stone deposit. The Koganezaki Crystal Park, a tourist destination that is famous for its glass craftworks, was founded featuring the Izu Silica Stone Mine. Glass is also one of the blessings of the volcano.

The period of terrestrial large volcanoes 87

A bird's-eye view of the form of the Koshimoda area. The gradual slope inclined downward toward the sea can clearly be seen. The picture was taken by the Izu Peninsula Geopark Promotion Council.

The remains of the Izu Silica Stone Mine in the Ugusu area, Nishi-Izu Town. The whole mountain looks white due to open-air mining.

37. Obsidian in Izu

In the previous section, I explained that Izu Silica Stone, from which glass is made, is a blessing of volcano. Silica stone itself is a white crumbly rock and just a glance at it may not make you think that it is a raw material for glass. However, a volcano sometimes directly erupts substances that can easily be recognized as glass even by layman's eyes. That kind of glass is called volcanic glass.

Obsidian is among such volcanic glass. Obsidian is semitransparent black or gray and has glassy gloss. It is completely similar to glass in how it breaks. It is difficult to imagine obsidian as a volcanic product from how it looks. However, it has the same element as rhyolite, which is a kind of volcanic rock. Obsidian spouted from the crater as lava flow or gushed out from the crater into the air in the form of rock fragments. However, obsidian is formed under particular conditions, such as the chilling of magma, and there are only limited places where it was created.

The places where a large amount of obsidian can be found are limited to southern Ikadaba in the city of Izu, the Kashiwatoge Pass on the border between the city of Izu and the city of Ito, Kamitaga in the city of Atami and Kajiya in Yugawara Town, Kanagawa prefecture. For southern Ikadaba in the city of Izu, the obsidian is part of the rhyolitic lava flows effused 3,200 years ago by Kawagodaira Volcano, which belongs to the Izu Tobu Volcano Group. The other ones excluding Izusan in the city of Atami whose geologic age is unclear are 300,000 to 600,000 years old. The rock form of the Kashiwatoge Pass is a rhyolitic lava dome 500 meters in diameter and part of it is obsidian. Probably, the obsidian in the other places, such as Kamitaga, is part of a small lava dome or lava flow.

Obsidian breaks in the same way as glass and is easy to make hard and sharp pieces. Those pieces were used as an arrowhead and a knife long time ago. Stone implements made of obsidian produced in Izu were discovered at historical sites in the Old Stone Age and the Jomon period in many parts of the Chubu and Kanto districts.

Rhyolitic lava that produced obsidian around Kamitaga and Kajiya was sometimes considered to be part of Yugawara Volcano and Taga Volcano mentioned in Section 31, but the rhyolitic lava is quite different in the nature of rock and distribution. I am tempted to include it in the Izu Tobu Volcano Group, but it is much older and even more eroded. It may be that there were small-scale activities of a monogenetic volcano group different from the Izu Tobu Volcano Group at the end of the era of

The period of terrestrial large volcanoes

terrestrial large volcanoes and consequently, rhyolitic lavas that produced obsidian were erupted.

An example of obsidian gleaming black. Its rugged surface is just like a broken glass. It was produced around the Kashiwatoge Pass in the city of Izu.

Obsidian ejected from Kawagodaira Volcano in the city of Izu. It can often be seen in rivers around the Ikadaba area in the city of Izu.

Chapter 4

The period of the Izu Tobu Volcano Group

38. A group of small volcanoes

As noted in Section 29, a volcano can be classified into polygenetic volcano or monogenetic volcano. On the Izu Peninsula after 150,000 years ago, large polygenetic volcanoes somehow had "died down" and monogenetic volcanoes, such as Omuroyama Volcano, began to erupt in many places. As a result, a group of small volcanoes (the Izu Tobu Volcano Group) was formed and the group has remained to this day. There are a total of nearly sixty volcanoes belonging to this group in the eastern half of the Izu Peninsula (the city of Izunokuni, the city of Izu, the city of Ito, Higashi-Izu Town and Kawazu Town) (see Frontispiece 16). Those volcanoes are also distributed on the sea floor between the Izu Peninsula and the Izu Oshima Island.

A monogenetic volcano can be classified into three types: scoria cone; tuff ring (or maar); or lava dome.

If splashes of magma (scoria) are erupted from the crater like a fountain, they pile up around the crater and create a scoria cone (1A). As seen with Omuroyama Volcano (refer to Sections 48 to 51) and Hachikuboyama Volcano (refer to Section 46), lava flows can spout from the foot of a scoria cone. If that happens, part of the scoria cone falls and gets on lava flows. This is called scoria raft (1B). In some cases, the fall of a scoria cone causes its shape to change (1C).

If magma meets a massive amount of underground water or seawater, it causes phreatomagmatic explosion (2A). A large dent of crater caused by it is called maar (2B). As seen with Lake Ippekiko in the city of Ito (refer to Section 40), water accumulates in a maar, which can create a lake. What is left as not only a large crater but also a ring-shaped mountain form surrounding it is called tuff ring (2C). As shown by Umenokidaira (refer to Section 41), some parts of ring-shaped landforms became U-shaped. As observed with Kadono and Ogi (refer to Section 42), a spout of lava flows at the end of eruption can cause the crater of tuff ring and maar to be buried or some part of it to overflow to the outside (2D).

If viscous lavas pile up around the crater, a lava dorm is formed (3A, 3B). One example of this is Yahazuyama Volcano (refer to Section 55). In addition, as seen with Omuroyama Volcano and Komuroyama Volcano, there are some plug domes that rose as if to put a cover on the spout because lavas got more viscous at the end of their outflow.

The period of the Izu Tobu Volcano Group

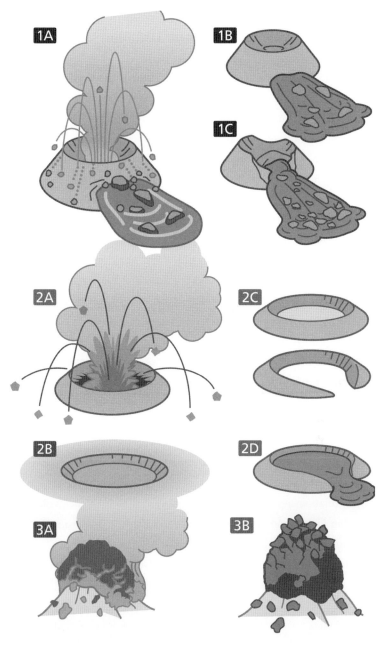

Types of monogenetic volcanoes and how they are formed (refer to the text). The illustrations were drawn by Sachiko Hagiwara.

39. Pumice and volcanic ash from Hakone

Hakone Volcano is an active volcano adjacent to the northern side of the Izu Peninsula. This volcano began to erupt 600,000 years ago. In particular, from 200,000 to 40,000 years ago, the volcano repeated explosive eruptions involving wide fallout of pumice and volcanic ash and large-scale pyroclastic flows.

Many of these volcanic products soared high into the skies along with volcanic smokes, were blown eastward by the prevailing westerlies and piled thickly on the Kanagawa prefecture on the lee side. Fortunately, because the Izu Peninsula is located south of Hakone Volcano, pumice and volcanic ash ejected from the volcano piled up only when rare northern winds blew.

Dozens of pieces of strata containing such pumice and volcanic ash were discovered on the Izu Peninsula and eight of them are distributed within the scope of the Izu Tobu Volcano Group. In ascending order, those strata are called Ohito Yellow Pumice 1 (128,000 years ago), Da-1 Pumice (125,000 years ago), Ohito Yellow Pumice 2 (117,000 years ago), Ohito Pink Pumice (110,000 years ago), Da-4 Pumice (100,000 years ago), Da-5 Pumice (88,000 years ago), TPfl ash (66,000 years ago) and Mishima Pumice (52,000 years ago). Some include particular geographical names and others include signs. Each of the layers has bright color tones, such as yellow, orange and pink, and can easily be distinguished from other volcanic ashes.

In particular, Da-1 Pumice, Da-4 Pumice, Da-5 Pumice and TPfl ash are distributed in a broad range of areas and can be found in many places, such as the city of Izunokuni, the city of Izu and the city of Ito. This is why it is easy to grasp their hierarchical relationships with volcanic ash originating from each volcano of the Izu Tobu Volcano Group and it was possible to use the Izu Tobu Volcano Group as the key bed to identify when those volcanic products were erupted.

In addition to Hakone Volcano, volcanic ash originating from more distant volcanoes was also discovered on the Izu Peninsula, although they are small in number. These examples are: Ontake Pm-1 Pumice that piled up by the major eruption of Ontake Volcano (3,067 meters above sea level) situated on the border between Nagano and Gifu prefectures 96,000 years ago; Kikai-Tozurahara Ash ejected from Kikai Caldera, a submarine volcano off the southern coast of Kagoshima prefecture, 92,000 years ago; and Aira-Tanzawa Ash ejected from Aira Caldera, a submarine volcano north of the Kagoshima Bay, 28,000 years ago. These volcanic ashes are difficult to discover because they are quite small in amount, but

The period of the Izu Tobu Volcano Group

they were very helpful to organize the eruptive history of the Izu Tobu Volcano Group.

Strata seen on the cliff around Ohito Town in the city of Izunokuni. Each of the brightly colored stripes is a layer of pumice and volcanic ash that piled up in Izu at the eruption of Hakone Volcano.

Strata of pumice originating from Hakone Volcano seen on the cliff around the Tawarano area in the city of Izunokuni

40. Lake Ippekiko and Numaike

Lake Ippekiko is a circular and beautiful lake 600 meters in diameter situated in the highland area between the Ito Spa and Omuroyama Volcano. A close look at the geomorphology around this area shows that on the southeastern side of Lake Ippekiko is a circular depression almost the same size. There is also a small lake on the northwestern side of this depression and this lake is connected with Lake Ippekiko through a narrow waterway (see Frontispiece 2 [below]).

These two concave areas are remnants of craters created by phreatomagmatic explosion. As evidence for this, there are many steep cliffs at the edge of the craters and you can see thick strata of explosion breccia containing many volcanic bombs. For the name of the volcano, the northwestern crater is called Ippekiko and the southeastern crater is called Numaike. Both of these are located just close to each other and there are no recognizable gaps in the period of their eruptions. Therefore, they can be considered a twin volcano formed by simultaneous eruptions.

However, although they are a twin volcano, there is no conspicuous landform other than the craters. This type of volcano is called maar, a word derived from German. There are many other examples of maar all over the world. Those examples in Japan are the port of Habunominato on the Izu Oshima Island, Ichinomegata on the Oga Peninsula and Lake Hangetsuko at the foot of Mt. Yoteizan in Hokkaido. In some cases, water accumulated in a crater, which created lakes or bays, and in other cases, only dried craters remained.

As I proceeded with my survey, I discovered a surprising fact. I found a unique orange pumice layer 40 centimeters thick sandwiched in between the explosion breccia that I mentioned earlier in this section. It was Da-4 Pumice that originated from Hakone Volcano, as mentioned in the previous section. This means that while Ippekiko and Numaike were erupting, there occurred a large-scale eruption of Hakone Volcano as well. It was an age of upheaval in terms of earth science. In addition, this fact also showed that Ippekiko and Numaike erupted 100,000 years ago as old as Da-4 Pumice.

The western end of Lake Ippekiko and the southern end of Numaike were partially filled in by the lava flows that were spouted from Omuroyama Volcano 4,000 years ago. A chain of small islands called Junirento Islands on the western end of Lake Ippekiko are a landform created by lava flows that were effused from Omuroyama Volcano and ran into the lake.

The period of the Izu Tobu Volcano Group

A bird's-eye view of Lake Ippekiko and Numaike. The Junirento Islands can be seen on the western side of Lake Ippekiko.

Explosion breccia that piled up around Lake Ippekiko after its eruption

41. Umenokidaira

To the southeast of Lake Ippekiko is a height called Umenokidaira. A topographical map shows that there is a U-shaped hill and its highest point is 297 meters. Around this hill are explosion breccias containing many volcanic bombs and it can be confirmed that this hill is a volcano. This volcano is called Umenokidaira Volcano after the place where it is located.

If you drive to the south along Route 135 from the Ito Spa, the road comes to an upward slope around past the Yoshida Basin and enters the northeastern edge of the crater of Umenokidaira Volcano. After the road runs across the crater field, it then gets out of the crater area over its southwestern edge. The diameter of the crater is 800 meters.

The nature of the eruption that created Umenokidaira Volcano is the same as the eruption that created the Ippekiko and Numaike maars, as noted in the previous section, and this eruption is called phreatomagmatic explosion, in which magma met a huge amount of underground water. However, Ippekiko and Numaike have no remarkable landforms except for the craters, whereas in the case of Umenokidaira, you can observe a ring-shaped swell of mountain form around its large crater.

This type of volcano is called tuff ring and is also broadly distributed all around the world just like maar. The tuff is a rock formed as a result of volcanic ash hardening. In some cases, certain parts of the ring-shaped mountain form are not conspicuous and they can sometimes become U-shaped or crescent like Umenokidaira Volcano. Probably, the most famous tuff ring in the world is Diamond Head located on the southeastern side of Waikiki Beach, Hawaii. A bird's-eye view of its ring-shaped mountain form is just like a diamond ring.

Like Ippekiko and Numaike Volcanoes, Da-4 Pumice originating from Hakone Volcano is also sandwiched in between the explosion breccia ejected from Umenokidaira Volcano. This means that Umenokidaira Volcano erupted and was created 100,000 years ago at the same time as adjacent Ippekiko and Numaike Volcanoes.

Umenokidaira Volcano spouted a massive amount of lava flows as well at the end of its eruption. These lava flows mostly moved eastward and poured into the Sagami Bay, which resulted in significantly expanding the land of the city of Ito. This expanded land includes the southern side of the Kawana Golf Course and the tableland of Sannohara to the south of the golf course.

The period of the Izu Tobu Volcano Group

A bird's-eye view of Umenokidaira Volcano seen from its northern side

Umenokidaira Volcano seen from the summit of Omuroyama Volcano. The round mountain on your left is Mt. Komuroyama and the flat hill occupying the right half space of the picture is Umenokidaira Volcano.

42. Kadono and Ogi

In the previous section, I explained that Ippekiko, Numaike and Umenokidaira Volcanoes to the south of Ito Spa had simultaneously erupted 100,000 years ago. These three volcanoes appear to be lined up in northwestern and southeastern directions. With a focus on the northwestern side of Lake Ippekiko, there is also a place where landforms created by volcanoes and volcanic products are complicatedly piled up on top of each other. As a result of in-depth onsite surveys, I was able to clarify the truth behind two more volcanoes.

The first thing to note is the lava plateau more than 50 meters in thickness on the eastern side of the Ito Okawa River. It seems that the valleys along the Ito Okawa River had been more open and spacious before lavas poured into them. It was Kadono Volcano that spouted the lavas.

The crescent hill (whose highest point is 217 meters) where the western half side of the resident area Kadonodai is situated is probably a tuff ring that forms part of Kadono Volcano. As mentioned in the previous section, a tuff ring is a ring-shaped or circular arc-shaped mountain that was created by phreatomagmatic explosion caused by a combination of magma and a large amount of underground water. It is conceivable that a massive amount of lavas overflowed from the crater on the western side of the tuff ring at the end of eruption and filled in the ravines along the Ito Okawa River.

Ogi Volcano is situated adjacent to the southeast of Kadono Volcano. A crescent hill on the northern side of a path running from Ito Spa to Lake Ippekiko can be considered part of its tuff ring. Another hill that seems to be part of the same tuff ring is also located around the resident area Ogi to the north of the former hill. It is thinkable that thick lavas filled in the crater with a diameter of 700 meters that used to be located between these two hills.

The five volcanoes of Ippekiko, Numaike, Umenokidaira, Kadono and Ogi form a volcano chain in northwestern and southeastern directions and can be considered to be created on the same eruptive fissure by simultaneous eruptions. Because all these five volcanoes caused phreatomagmatic explosions, their volcanic ash piled up in a vast range of eastern Izu with a thickness of more than 50 centimeters around Sakuranosato on the western side of Omuroyama Volcano.

The period of the Izu Tobu Volcano Group

The cliff 150 meters high surrounding a DIY store along the Ito Okawa River is made of a thick lava flow that filled in the crater of Kadono Volcano.

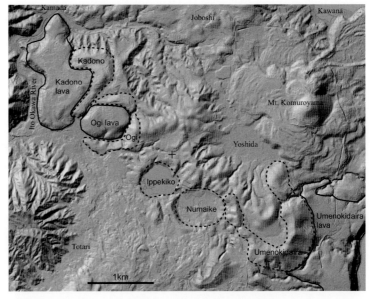

A 3-D topographic map of Kadono, Ogi, Ippekiko, Numaike and Umenokidaira Volcanoes. Broken line shows an outline of each tuff ring or maar. Solid line shows an outline of each lava flow. The map was drawn from the Fundamental Geospatial Data developed by the Geospatial Information Authority of Japan.

43. What a volcano chain means

Volcano chains (or crater chains) in which small volcanoes and craters, including examples of the Izu Tobu Volcano Group, are lined up in a straight line can be observed everywhere. For example, when Izu Oshima Volcano erupted in 1986, 20 craters opened in a range from around the summit of Mt. Miharayama to its northwestern side and formed a clear crater chain.

Many observational facts demonstrate that just under these crater chains is a linear fissure and that magma rose along the fissure and caused an eruption. In fact, there are cases in which dikes formed by cooled and hardened platy magma can be actually observed inside of craters. I guess that people imagine cylindrical pipes as where magma runs, but as a matter of fact, such cylindrical pipes exist only underground in places where eruptions were repeated over a long period of time, such as the crater at the summit of Fuji Volcano.

Other places where magma runs are platy almost unexceptionally and eruptions occur at some ground points that magma reaches, which results in forming volcano chains (or crater chains). This is because magma can create platy fissures more easily than cylindrical pipes with even less workload.

Those fissures find the easiest direction in which they can open. The Izu Peninsula is pushed to the Japanese main island by the northwestern movement of the Philippine Sea Plate and its crust is under heavy pressure in northwestern and southeastern directions. It is the easiest way in which magma creates dikes along with this direction and pushes the crust open in a direction perpendicular to them (that is, northeastern and southwestern directions). Because of this situation, many volcano chains of the Izu Tobu Volcano Group stretch in northwestern and southeastern directions (see Frontispiece 15).

The period of the Izu Tobu Volcano Group

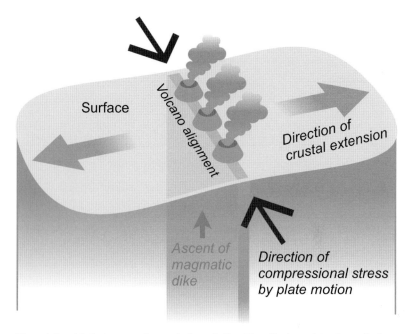

The relationship between volcano chain and dike. The dike is a platy channel where magma runs. The illustration also shows the relationship between volcano chain, dike and the direction of force that works on the crust. The illustration was drawn by Sachiko Hagiwara.

A group of dikes seen on the inner wall of the Hoei Crater of Fuji Volcano. It is considered that these dikes also lie under the volcano chain of the Izu Tobu Volcano Group.

44. The largest eruption of Hakone Volcano

As noted in Section 39, TPfl volcanic ash piled up in even Izu by the eruption of Hakone Volcano approximately 66,000 years ago. In this section, I will focus on the dreadful truth behind the eruption. To tell the truth, it is not so appropriate to say that this volcanic ash "piled up." This volcanic ash is the remnant of a pyroclastic flow that was ejected and ran down from Hakone Volcano. TP stands for *Tokyo Pumice* in English and fl represents *flow* in English. A pyroclastic flow is the phenomenon in which the smoke of a volcano runs and spreads at a quick pace as if it crept over the earth surface. The real substance of it is a high-temperature gas containing volcanic ash and small stone.

Hakone Volcano had caused pyroclastic flows many times long before that. However, those pyroclastic flows did not reach so distant a place and just stopped in the city of Mishima and Kannami Town on the southern side of the volcano. However, the eruption of Hakone Volcano 66,000 years ago discharged magma with a volume of even five billion cubic meters. In comparison with magma with a volume of 300 million cubic meters spouted by the largest eruption of the Izu Tobu Volcano Group in history, it is self-evidently clear how overwhelming the eruption of Hakone Volcano was 66,000 years ago.

This eruption initially spouted a high-rising eruption column that reached the stratosphere over the crater and caused a hail of pumice to be piled up in Tokyo, as shown by its name "Tokyo Pumice." These piles of pumice can still be observed in Tokyo in the form of strata 20 centimeters thick. However, the nature of the eruption dramatically changed immediately after that. Volcanic smokes that soared high into the skies lost buoyancy for some unclear reason and suffered gravitational collapse. However, those volcanic smokes contained much air and were fluidized. Because of this, collapsed volcanic smokes began to spread in every direction as pyroclastic flow, remaining in a high temperature.

Consequently, the pyroclastic flow covered a vast expanse of areas ranging from around the mouth of the Fujikawa River to the west to Totsuka in Yokohama to the east. All creatures living along these routes were burned to death. In addition, the pyroclastic flows reached the Kanogawa River area in the city of Izu to the south and the tableland on the southern side of the Ito Spa. The strata formed by the pyroclastic flow well show the characteristics of flow deposits. Specifically, the strata contain both fine volcanic ash and rock blocks. Volcanic ash that drifted in the air and piled

The period of the Izu Tobu Volcano Group

up on the grounds do not have such characteristics because its particles become similar in size. After this enormous eruption, the activitiy of Hakone Volcano became less active and the volcano seldom had an eruption that even caused damage to Izu.

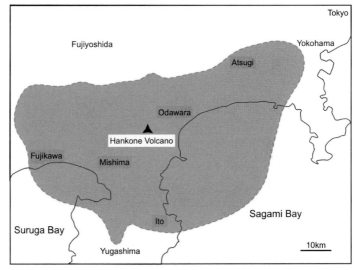

The range of areas covered with the TPfl pyroclastic flow caused by the eruption of Hakone Volcano approximately 66,000 years ago

The thick rocky layer in the middle of the photo is the pyroclastic flow (TPfl) caused by the tremendous eruption of Hakone Volcano approximately 66,000 years ago. They contain boulder-sized rocks. The picture was taken in the Shimonagakubo area, Nagaizumi Town.

45. A volcano that created the Kawazu Nanadaru Falls

Two kilometers southeast of the Amagitoge Pass is Mt. Noborio (1,057 meters above sea level). Topographically, it is proper to say that it is a ridge protruding southwestward from Mt. Amagisan and geologically, it is part of Amagi Volcano (refer to Section 32). There was an eruption on the southern slope (around 700 meters above sea level) of Mt. Noborio approximately 25,000 years ago. This marked the birth of Noborio Minami Volcano of the Izu Tobu Volcano Group.

This eruption occurred on a steep slope and the form of the volcano itself is unclear. However, particularly notable is a lava flow that came out of there. The lava flow ran down 1.5 kilometers west-southwestward on the slope of Mt. Noborio and came to the Kawazugawa River. Then, the lava flow changed their direction to the southeast along the river and streamed further two kilometers filling in the valley. The surface and inside of the lava flow washed by water-stream erosion produced the smooth bedrock seen everywhere on the riverbed of the Kawazugawa River within this range.

A close examination shows that columnar joints can be observed on the surface of the bedrock. Columnar joints are prism-shaped fissures that are formed when melt rocks cool and harden and their volumes contract. The cross sections of those rectangular columns are often hexagonal, but some are pentagonal and heptagonal. This rectangular column is apt to stretch in a direction in which it is deprived of heat.

Seven falls are running onto the steps on the bedrock and this is the famous Kawazu Nanadaru Falls. From upstream, they are called the Kamadaru Falls, the Ebidaru Falls, the Hebidaru Falls, the Shokeidaru Falls, the Kanidaru Falls, the Deaidaru Falls and the Odaru Falls. However, I have recently discovered that one of these seven falls has nothing to do with the bedrock made of lava flows. I will leave the joy of finding which fall is the one I have just mentioned to my readers. In this connection, the bedrock formed by lava flows can also be observed in Sarutafuchi, a little upstream the Kamadaru Falls.

There are almost uncountable falls in the mountains in Izu. Probably, the reason why the Kawazu Nanadaru Falls gained the status of a major tourist destination of all these numerous falls is that in addition to their geographical advantage of belong located along the streets through the Amagitoge Pass, their rocks are notably fresh because of their geologic youth of being 25,000 years old and their beautiful bedrock with columnar joints is spectacular.

The period of the Izu Tobu Volcano Group

The Odaru Falls with the largest drop of the Kawazu Nanadaru Falls

The Kamadaru Falls, which run the most upstream of the Kawazu Nanadaru Falls. The picture was taken by Yusuke Suzuki.

46. Hachikuboyama and the Jorennotaki Falls

The Kanogawa River, which has the largest size of a catchment basin and is also the longest in Izu, branches into two rivers with different names up the Yugashima Spa in the city of Izu: the Nekko River on the western side and the Hontanigawa River on the southern side. Route 414 runs southward along the Hontanigawa River and leads to the Amagitoge Pass. Halfway through this route, the famous Jorennotaki Falls runs two kilometers upstream the Yugashima Spa (see Frontispiece 7 [above]). Spectacular columnar joints can be observed on the bedrock forming the fall, which demonstrates that this fall was created by a lava flow.

The lava flow was gushed out when Hachikuboyama Volcano (674 meters above sea level), which is situated one kilometer southeast of the falls, erupted 17,000 years ago. The lava flow created a gradual tableland around the Jorennotaki Falls and on this tableland are tourist facilities and parking lots of the falls and the Kayano community in the city of Izu and its farmland (see Frontispiece 6). On a topographical map, this tableland stretches northward to the Yugashima Spa in a tapering form between the Hontanigawa River and its branch Iwabigawa River to the east of it. This form of the tableland is exactly the landform created by the lava flow that used to run filling in the ravines. Fortunately, the lava flow stopped one kilometer ahead of the Yugashima Spa. In addition, if the amount of erupted lava flows had been larger, it would have reached the Yugashima Spa and this spa resort, which boasts of beautiful ravines aroud it, may not have been able to gain its present scenic beauty.

Hachikuboyama Volcano is a scoria cone that was formed by piles of low-viscosity splashes of magma (scoria) around the crater. Its bottom is 800 meters in diameter and it is less than 300 meters in height. The volcano is pudding-shaped. However, unfortunately, it was created in steep mountains and only a few people know what a beautiful form it has.

There is a round hill (938 meters above sea level) called Mt. Maruyama 1,200 meters southeast of Hachikuboyama Volcano. This hill is also a scoria cone and is considered to have erupted at the same time as Hachikuboyama Volcano. Like other volcano chains of the Izu Tobu Volcano Group, both Hachikuboyama Volcano and Maruyama Volcano is a volcano chain formed by fissure eruption in northwestern and southeastern directions.

The period of the Izu Tobu Volcano Group

Hachikuboyama Volcano seen from its northwestern side

The cross section of the Maruyama scoria cone seen along a woodland path. The scoria cone contains many volcanic bombs.

47. The eruption of Fuji Volcano and Izu

Mishima, located at the northern edge of the Izu Peninsula, is surrounded by mountains in three directions: Mt. Ashitakayama on its northwestern side; Mt. Hakone on its eastern side; and the Shizuura Mts. on its southern side. In addition, to the northwest of Mt. Ashitakayama Volcano is active Fuji Volcano. Fuji Volcano began to erupt approximately 100,000 years ago, which means that its whole period of activity overlaps with the active period of the Izu Tobu Volcano Group (from approximately 150,000 years ago to the present).

The Shizuura Mts. is old mountains formed by submarine volcanoes belonging to the Shirahama Group, which I explained in Sections 12 to 18, and has a complicated shape under the influence of erosion. Ashitaka Volcano and Hakone Volcano, which formed Mt. Ashitakayama and Mt. Hakone respectively, began to erupt approximately 400,000 to 600,000 years ago. However, Ashitaka Volcano stopped erupting 100,000 years ago and has many deep ravines created by subsequent erosions. Hakone Volcano is an active volcano that still continues its activity. However, the last time its eruption affected even the landform around Mishima was the enormous eruption 66,000 years ago, as noted in Section 44. Many of hills lying on the eastern side of Mishima were formed by thick piles of the pyroclastic flow ejected by this huge eruption.

Between Ashitaka Volcano and Hakone Volcano lies a wide ravine through which the Kisegawa River and the Daibagawa River run. However, both rivers are not in the center of the ravine. The Kisegawa River runs along the foot of Ashitaka Volcano and the Daibagawa River runs along the foot of Hakone Volcano. Fundamentally, it is unnatural that two different rivers run through a ravine in line with each other without meeting.

More strange is the route the Daibagawa River takes around Mishima. The Kisegawa River directly runs into the Kanogawa River along the shortest route. In contrast, the Daibagawa River widely bypasses the urban districts in Mishima, flowing to its border with Kannami Town to the far south and then running into the Kanogawa River.

This unnatural route of river flows reflects a small difference of elevation in valleys and plains. The urban districts in Susono and Mishima form tablelands higher than their surrounding areas and the above-mentioned two rivers avoid running across the tablelands. In addition, these tablelands were created by massive lava flows

The period of the Izu Tobu Volcano Group

(Mishima Lava) effused from Fuji Volcano approximately 10,000 years ago. These lava flows can be seen as layers of black rock on the riverbed of the Kisegawa River and everywhere in the urban districts in Mishima. Snow water from Fuji Volcano running through the fine fissures in the lavas is gushing out in a steady stream from the end of lava flows. Examples of this are Kohamaike Pond and Komoike Pond in the urban districts in Mishima and the Kakitagawa River in Shimizu Town.

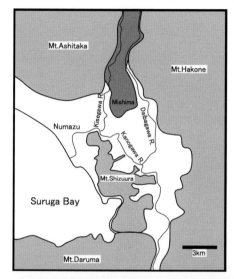

The landform around Mishima and the route of river flows. The light gray parts show mountainous areas and the dark gray parts show the rough distribution of Mishima Lava effused from Fuji Volcano.

Fuji Volcano seen from the Kanogawa River, which runs through the northern part of the Izu Peninsula

48. Omuroyama (1): Scoria cone and crater

Needless to say, Mt. Omuroyama (580 meters above sea level), which rises at the highest point of the Izukogen Plateau, is the most significant landmark of the city of Ito and is also an iconic mountain that can be said to be the symbol of the city.

In volcanological terms, Mt. Omuroyama is the largest scoria cone (Omuroyama Volcano) of the Izu Tobu Volcano Group. If low-viscosity magma spouts from the crater like a fountain, it immediately chills and hardens, turning into a dark colored pumice-like rock. This is a scoria and the mountain created by piles of fallen scoria around the crater is a scoria cone. If a scoria cone grows higher as eruption progresses, fallen scoria rolls down the slope unstably. The scoria that rolls down into the inside of the crater's rim returns to the bottom of the crater and repeatedly spouts out again. The scoria that rolls down into the outside of the crater's rim causes the foot of a scoria cone to gradually grow. In this way, a mountain form like a huge pudding was finally created with its bottom 1,000 meters in diameter and 300 meters in height from the bottom (see Frontispiece 2 [above]).

At the summit of Omuroyama Volcano is an old conic crater 250 meters in diameter and 40 meters in depth. Lavas accumulated inside this crater at the final stage of eruption and a lava lake was created in it. A small shrine called Sengen Shrine was built on the northeastern side of the crater's inner wall as if it were clinging to it. A dark gray slab of volcanic rock can be seen on the cliff at the back of this shrine and this rock is part of the lava lake that used to fill in the crater. Most of the lavas in the lava lake later disappeared, returning to the underground and overflowing to its surrounding spaces, but only the lavas sticking to the crater's inner wall remained. A close observation of the rock surface enables you to find volcanic bombs that dropped into the lava lake and came near to becoming part of it when the volcano erupted.

In the meantime, a careful person would be able to find a small crater 50 meters in diameter at a spot around 450 meters above sea level on the southern slope of Omuroyama Volcano as well (see Frontispiece 3 [above]). Because the shape of this crater is clear, it can be considered to have been formed after Omuroyama Volcano almost finished growing. Probably, on the final phase of eruption, it was difficult for the pressure on underground volcanic gas to escape to the crater at the summit and it sought a passage of escape in other places. As a result, the gas slightly overflowed to the sides, which created a small crater.

The period of the Izu Tobu Volcano Group

A bird's-eye view of Omuroyama Volcano seen from its western side. Its pudding-like shape is beautiful.

The inside of the crater at the summit of Omuroyama Volcano. The structure slightly to the right from the middle is Sengen Shrine.

49. Omuroyama (2): Lava outlet

Omuroyama Volcano spouted a massive amount of lava. Roughly calculated, the erupted lavas amounted to overwhelming 380 million tons, worth approximately 100 million four-ton trucks. This enormous amount of lavas filled in the bumpy landform that used to be around the mountain, which resulted in creating the gradual Izukogen Plateau.

From what part of Omuroyama Volcano did these lava flows erupt? To the northeast of Mt. Omuroyama sits Mt. Iwamuroyama (448 meters above sea level). Mt. Iwamuroyama is approximately 500 meters in diameter and 70 meters in height from its foot. On its flat summit is the Izu Shaboten Park, but its mountain side is steep and the whole mountain is dome-shaped. The summit and mountain sides are filled with large angular rocks. These characteristics show that Mt. Iwamuroyama is a lava dome. A lava dome is the same type of volcano as Mt. Heisei Shinzan of Unzendake Volcano and is created when viscous lavas swell from the crater during the 1991-1996 eruption. To the south of Mt. Omuroyama is also a mountain (310 meters above sea level) quite similar to Mt. Iwamuroyama and it is called Mt. Moriyama. Mt. Moriyama is smaller than Mt. Iwamuroyama and inconspicuous, but the mountain is also a lava dome.

If you trace the lavas around Omuroyama Volcano upstream, you will find that many of them run from Mt. Iwamuroyama and Mt. Moriyama. That is, these lava flows did not directly come from Mt. Omuroyama but from the outlet on its foot. Mt. Iwamuroyama and Mt. Moriyama are plug domes that rose as if they put a lid on the outlet when the lavas got more viscous near the end of eruption (see Frontispiece 3 [above]).

The eruption of Omuroyama Volcano not only spouted 380 million tons of lava flow but also caused a total of 130 million tons of volcanic ash to pile up around the mountain (see Frontispiece 7 [below]). These piles of volcanic ash are 50 centimeters thick even around Lake Ippekiko three kilometers away from Omuroyama Volcano.

Recently, trees buried under volcanic ash ejected from Omuroyama Volcano have been discovered at a construction site in the Izukogen Plateau and their radiocarbon age was measured at approximately 4,000 years old. In addition, almost the same age has also been estimated from the relationship between the Jomon earthenware and volcanic ash at a historical site around the Ito Spa. Based on these data, Omuroyama Volcano is considered to have erupted approximately 4,000 years ago.

The period of the Izu Tobu Volcano Group

A bird's-eye view of Omuroyama Volcano seen from the eastern side

Omuroyama Volcano, the Izukogen Plateau and the Jogasaki Coast seen from their southern side. You can clearly see that the lavas effused from Omuroyama Volcano created a gradual slope (the Izukogen Plateau) and reached the Jogasaki Coast.

50. Omuroyama (3): A dammed lake and lava coast

The first lava flow that gushed from the western foot of Omuroyama Volcano amounted to 13 million tons. The lava flows branched into northward and southward streams and ran down to Totari and Ike in the city of Ito, respectively. At that time, the landform around Ike was totally different from what it is today and there is a deep ravine stretching from around the Rokurobatoge Pass on the western side of Ike. The lava flows poured into this ravine and dammed up halfway through it. As a result, the river running through the ravine was blocked, which created a lake that was probably twice as large as Lake Ippekiko. This lake is the origin of the current geographical name of Ike (pond in English). The lake was gradually filled in by rock debris coming from the surrounding mountain and had contracted to even one third of what it initially was by 1868. It is the current Ike Basin that was created through reclamation by digging a spillway tunnel at the exit of this lake in 1869.

Next, nearly 400 million tons of lava flow that exceeded the first lava flow by thirty-fold erupted from two points (Mt. Iwamuroyama and Mt. Moriyama mentioned in Section 49) at the northeastern and southern foots of Omuroyama Volcano and began to run in northern, eastern and southeastern directions. The lava flows that headed northward passed by the western side of Lake Ippekiko and poured into the Ito Okawa River at a southeastern point of Mt. Shiroyama. Then, the lava flows drifted one kilometer northward and reached around the Kamada area in the city of Ito. A small part of the lava flows streamed into Lake Ippekiko and created a chain of islands called the Junirento Islands. The lava flows also got into the crater of Numaike Volcano on the southeastern side of Lake Ippekiko and buried the southern half side of the crater. In addition, the lava flows interrupted the ravine around Totari, which resulted in creating another lake. This lake was subsequently filled in naturally and became what is now the Totari Basin.

The lava flows that streamed eastward from Omuroyama Volcano ran down the ravine from the northern side of Izu Granpal Park to the east, reaching the sea past around Futo Station on the Izu Kyuko Railway Line and created a lava fan on the coast (see the photo in Section 83). The town that was formed on this fan is the current spa area of Futo. The lava flows that headed southeastward were the largest in amount and poured into the Sagami Bay in a range with a width of four kilometers between Harai and Yawatano in the city of Ito. This filled in the sea and put the coastline forward by nearly two kilometers at some spots. The eruption created an

The period of the Izu Tobu Volcano Group

additional broad range of land. The place at the tip of these lava flows corresponds to what is the Jogasaki Coast now (see Frontispiece 1). It is conceivable that the coastline before the eruption of Omuroyama Volcano was situated between the current Route 135 and the Izu Kyuko Railway Line.

The Ike Basin seen from the summit of Mt. Omuroyama. Under the town on the left (eastern side) of the basin are lava flows effused by the eruption of Omuroyama Volcano.

The Jogasaki Coast, which was created after the lava flows from Omuroyama Volcano had filled in the sea

51. Omuroyama (4): Pothole and scoria raft

The eruption of Omuroyama Volcano created various natural landforms and products. In this section, I will introduce two major examples. There is a strange-shaped rock like a half cut of the body of a capsized ship, protruding from the grass of the Sakuranosato Park at the northwestern foot of Omuroyama Volcano. The bottom of the ship is made of a slab of lava rocks and is crammed with rough and dry reddish black pumice (scoria). This is called scoria raft and used to be part of Omuroyama Volcano. When a lava began to spout from the western foot of Omuroyama Volcano, the lava destroyed part of the volcano just above it and made scoria rafts. Because scoria includes so many bubbles and is so light that it did not sink into the lava flows. Instead, scoria rode on the lava flows and rolled on their surface. In this process, the lavas stuck to the surface of scoria. Scoria rafts can be seen not only in the Sakuranosato Park but also on the cliff at a construction site in the Izukogen Plateau.

There is a place called the Kannonhama Beach in the middle of the Jogasaki Coast, which was created by the lava flows from Omuroyama Volcano and you can observe a naturally created object called pothole. Imagine that a large rock is placed on a slab of hard bedrock for some reason. What will happen if that place is the inside of a river or the water's edge? Under such circumstances, the big rock would be constantly moving along with river flows and waves and would gradually dig a hole through its bed. In this process, the big rock itself would also gradually be worn out and eventually disappear with the result of only the bedrock with a round hole remaining. This round hole is called pothole. Pothole itself is not so rare. However, the pothole on the Kannonhama Beach still remains in the process of a big rock digging a hole. This big rock itself originally used to be part of lava flows, but it completely lost its corners and totally became a ball. Its surface has been polished like a mirror (see Frontispiece 4 [below]). This type of pothole is extremely rare. Probably, the rock took hundreds of years to become a ball. Currently, it is designated as a natural monument by the Ito municipal government, but it is quite reasonable that the pothole be upgraded to a central government-designated natural monument. I sincerely hope that such miraculous natural product will be preserved with great care and attention as long as possible.

The period of the Izu Tobu Volcano Group

A scoria raft at the western foot of Omuroyama Volcano

A pothole on the Kannonhama Beach along the Jogasaki Coast

52. Kawagodaira (1): Lava flow and pyroclastic flow

There is a one-kilometer east-west stretch of depression called Kawagodaira just on the northern side of the ridge line 1,500 meters west of Banzaburodake (1,405 meters above sea level), which is the highest peak of Mt. Amagisan renowned for its forest of rhododendrons. This depression has a U-shaped landform that opens to the north and its bottom is 1,090 meters above sea level. The depression is a huge crater created by explosive eruption and is called Kawagodaira Volcano after the geographical name of the place where it is located.

Its geomorphology clearly shows that thick lavas gushed northward from this crater (see Frontispiece 5). The lava flows were 50 meters thick. They ran down four kilometers along the northern slope of Mt. Amagisan and reached the southern side of Ikadaba in the city of Izu. The end of the lava flows and their right and left edges form steep cliffs, which shows that the lava flows were viscous.

The rock forming these lava flows is a volcanic rock called rhyolite, which is rare for the Izu Tobu Volcano Group and looks like a grayish white rugged rock. However, it is very light like pumice because it contains many bubbles from which volcanic gas escaped. Since a long time ago, people have paid attention to its light weight and thermal resistance and have used the rock as building material.

Something further extraordinary is distributed from the northern end of the lava flows to the area around Iakadaba—thick pyroclastic flow. A pyroclastic flow creates unique-looking strata when a mixture of high-temperature volcanic gas and volcanic ash runs down the mountain slope at a quick pace and gas escapes after it has stopped. The pyroclastic flows distributed around Iakadaba form strata in which grayish white fine volcanic ash with a touch of pink is dotted with white pumice and they are even 20 meters thick.

Strata created by pyroclastic flow are often dotted with black charcoal. This is because the plants grown there in time of volcanic eruption were burned by the heat of pyroclastic flow and was carbonized. In some cases, strata of pyroclastic flow contain parts of thick boughs and trunks and a whole huge tree. Such huge trees are called Jindai (mythological age) cedar or Jindai hinoki cypress and their inner elements that are not carbonized were sometimes used as wood materials. You can see specimens of Jindai cedar or Jindai hinoki cypress at the Naka Izu Museum of History and Folklore in the Kamishiraiwa area, the city of Izu, and the Izu Peninsula Geopark Amagi Visitor Center in the Yugashima area, the city of Izu.

The period of the Izu Tobu Volcano Group

The cross section of the lava flows effused from Kawagodaira Volcano

A steep ravine running through a tableland formed by the pyroclastic flows ejected from Kawagodaira Volcano

53. Kawagodaira (2): The entire process of eruption

Based on the studies made so far, the eruption process of Kawagodaira Volcano is roughly considered to have been as follows. Ominous earthquake swarms began underground Mt. Amagisan on a day 3,200 years ago. Subsequently, around summer, a crater finally opened and a series of explosive eruptions began and a huge mushroom-shaped eruption column rose to the stratosphere. Normally, a western wind is blowing in the lower stratosphere, but this western wind weakens in the summertime and winds of different directions sometimes blow under the influence of various atmospheric phenomena. The eruption column was blown by the southeastern wind in that season to the northwestern side of the crater and hails of pumice and volcanic ash carried by the eruption column fell down on the areas on the lee side. This pumice and volcanic ash are thickly distributed in a northwestern direction from the crater and have been discovered at the foot of Fuji Volcano, in the Shizuoka Plains, at the bottom of Lake Hamanakako and even in the Hira Mountains on the western side of Lake Biwako, Shiga prefecture.

Subsequently, the eruption entered on a more serious phase—the occurrence of pyroclastic flows. Part of the eruption column collapsed and began to travel down the slope of Mt. Amagisan in the form of pyroclastic flows. The first two pyroclastic flows drifted northwestward and their forefront reached even around the Yugashima area in the city of Izu. Following this, there were repeated downfalls of pumice from the eruption column and small-scale pyroclastic flows. The last pyroclastic flow was the largest in scale. Thirty million tons of magma flowed at a time in the form of pyroclastic flows. Those flows swept across the area around Ikadaba in the city of Izu and also ran to Higashi-Izu Town beyond the ridge line of Mt. Amagisan. In this connection, the wind direction until this point of time was southeastward, unchanged from when the eruption had started, which means that things happened for so short a time as not to allow a change in the wind direction, probably within a few days.

Subsequently, a huge amount of magma from which gas escaped was effused from the crater and 400 million tons of lava slowly streamed down northward, which marked the end of all eruptions. The amount of magma spouted by this eruption totaled even 760 million tons, which was the largest of the Izu Tobu Volcano Group. The eruption destroyed a vast expanse of forest. Because of this, every time it rained after the eruption had stopped, large-scale debris flows occurred and ran down the Omigawa River many times.

The period of the Izu Tobu Volcano Group

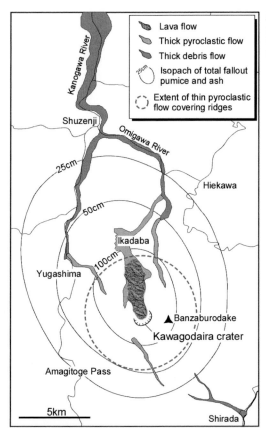

A mapped distribution of the main volcanic products from Kawagodaira Volcano. The thin solid lines show major roads.

A pile of pumice and volcanic ash ejected from Kawagodaira Volcano

54. Kawagodaira (3): Horrible eruption

I guess that the readers who have carefully followed this book so far have already realized one thing: Kawagodaira Volcano has several unique characteristics that other volcanoes of the Izu Tobu Volcano Group did not have when they had erupted before Kawagodaira.

Firstly, rhyolitic magma erupted for the first time in the eruptive history of the Izu Tobu Volcano Group. All of the magma that had been ejected from the Izu Tobu Volcano Group before Kawagodaira Volcano was relatively low-viscosity basaltic and andesitic. Low-viscosity magma is easier for volcanic gas to escape from, which makes it less likely that explosive eruptions due to a rise in gas pressure will occur. In contrast, because rhyolitic magma is of high viscosity, it can cause explosive eruptions more easily and requires outstanding caution in terms of disaster damage control.

Secondly, pyroclastic flows erupted for the first time in the eruptive history of the Izu Tobu Volcano Group. As noted in Section 44, pyroclastic flows are a dangerous phenomenon in which high-temperature and high-speed eruption columns travel a long distance along the surface of the ground and there is no other choice but to quickly escape to a faraway place in terms of disaster damage control.

Thirdly, a vast expanse of area was damaged. The eruption of Kawagodaira Volcano caused downfalls of a massive amount of volcanic ash and pumice in a broad range of Izu and destroyed forests. Even afterward, every time it rained, large-scale debris flows occurred and even affected the area around the mouth of the Kanogawa River. Downfalls of volcanic ash damaged not only Izu but also a wide expanse of area in the Chubu district.

As these show, the eruption of Kawagodaira Volcano was an extraordinary one that included many first-time factors in the eruptive history of the Izu Tobu Volcano Group. Ordinarily, preventive measures for volcanic disasters are formulated on the basis of eruptive history. Therefore, if phenomena that have never occurred before in eruptive history occur one after another, experts may be forced to throw their hands up in despair in planning disaster damage control measures. Now that we have already known that such a terrible eruption as Kawagodaira Volcano can occur in Izu, we can make preparations for it. However, suppose that now were the time immediately before the eruption of Kawagodaira Volcano 3,200 years ago, it means that we would soon be going to experience a completely unpredictable level of

The period of the Izu Tobu Volcano Group

eruption. The mere thought of it sends a chill of fear down my spine. The eruption of Kawagodaira Volcano gave us significant lessons that volcanic disasters can cause totally unpredictable things and that we need to put in place the minimum preventive measures to avoid falling into the desperate situation in which we have no other choice but to throw our hands up in despair if such unpredictable things occur.

▲The strata just above the pickax were created by the eruption of Kawagodaira Volcano. The striped strata on the lower layer are piles of pumice and volcanic ash falled from eruption columns. The stripe-free layer just above them is a pyroclastic flow ejected from the crater. The picture was taken around the Jizodo area, the city of Izu.

◀ A tree's trunk involved in and burned by a pyroclastic flow ejected from Kawagodaira Volcano. Its vertical erection means that it was burned while standing.

55. Iwanoyama- Ioyama volcano chain

If you look to the west from the summit of Mt. Omuroyama, you will find a rugged dome-shape mountain protruding from halfway up Mt. Amagisan. It is Mt. Yahazuyama (816 meters above sea level), which is known as "Genkotsu (Fist) Mountain" among the citizens of Ito (see the photo in Section 50). Mt. Yahazuyama is a lava dome formed after high-viscosity lavas swelled from the crater. A close observation shows that there is also a similar-shaped low mountain to the northwest of Mt. Yahazuyama. This mountain is also a lava dome and is called Mt. Ananoyama (660 meters above sea level).

The term lava dome became well-known after Unzendake Volcano in Nagasaki had begun to erupt in 1990. The lava dome created by this eruption is what is Mt. Heisei Shinzan today. In the case of lava domes abroad, Puy de Dome (in the mid-part of France), which is depicted on the label of the mineral water "Volvic", is particularly famous.

A lava dome is a rare case for the Izu Tobu Volcano Group. There are only four examples, including Mt. Yahazuyama and Mt. Ananoyama mentioned in this section (There are just a total of seven examples, even including special cases in which lava domes form at the outlet of lavas, as mentioned in Section 49). This is because most of the lavas effused from the Izu Tobu Volcano Group are of low-viscosity, which makes it relatively difficult for lava domes to be formed.

There is another lava dome called Mt. Iwanoyama two kilometers northwest of Mt. Ananoyama. Between Mt. Ananoyama and Mt. Iwanoyama are four small depressions and each of them, seen from northwest, is called Iwanokubonishi, Iwanokuboshigashi, Fujimikubo and Ananokubo. These depressions can be considered to be craters that caused small-scale phreatic explosions.

In the meantime, if you look in the opposite direction, you will find that three kilometers to the southeast of Mt. Yahazuyama is Mt. Ioyama (459 meters above sea level). Mt. Ioyama is a hill behind the Akazawa retreat area in the city of Ito. Like Mt. Omuroyama, it is a scoria cone formed after splashes of low-viscosity magma had piled up around the crater. As much as 200 million tons of lava were effused from Ioyama Volcano and poured into the Sagami Bay to the east of it, which created a spacious tableland on which the current Ukiyama Spa sits (see Frontispiece 3 [below]).

These above-mentioned volcanoes and craters form a spectacular northwest-

The period of the Izu Tobu Volcano Group

southeast volcano chain and the distance from one end to the other is approximately six kilometers. As this book has already discussed several similar cases, this volcano chain is considered to have been created after it simultaneously erupted on the same eruptive fissure. The eruption occurred approximately 2,700 years ago.

Mt. Yahazuyama (left) and Mt. Ananoyama (right) seen from around the Ike Basin in the city of Ito.

A bird's-eye view of Mt. Ioyama seen from its eastern side. Two streams of lava effused from Mt. Ioyama poured into the Sagami Bay at the front and created the two tablelands of lava on which the Ukiyama Spa sits.

56. A summary of eruptive history (1): Locations of eruption and magma types

In this section, I will look back on the whole history of the terrestrial areas of the Izu Tobu Volcano Group with a primary focus on eruption spots. From 150,000 to 80,000 years ago, volcanoes north of Togasayama Volcano erupted one after another. Those volcanoes include the Takatsukayama-Sukumoyama volcano chain (131,000 years ago) that straddles the city of Izunokuni, the city of Izu and the city of Ito, Hinata Volcano (129,000 years ago) and Funabara Volcano (150,000 years ago) in the city of Izu and the Kadono-Umenokidaira volcano chain (100,000 years ago) in the city of Ito. That is, it can be said that the northern half of the Izu Tobu Volcano Group was created during this period of time.

Subsequently, from 80,000 to 20,000 years ago, in addition to the continued eruption of volcanoes in the city of Ito, it is notably significant that many volcanoes erupted in the southern part of the city of Izu and in Kawazu Town. Those volcanoes include Hachinoyama Volcano (36,000 years ago) and Noborio Minami Volcano (25,000 years ago) in Kawazu Town. That is, the activity area of the Izu Tobu Volcano Group spread to the south during this period of time.

In addition, from 20,000 years ago to the present, volcanic eruptions have been active in the whole extent of the Izu Tobu Volcano Group except for the area of the city of Izunokuni. Those volcanoes include Komuroyama Volcano (15,000 years ago), Omuroyama Volcano (4,000 years ago), the Iwanoyama-Ioyama volcano chain (2,700 years ago) in the the city of Ito, Hachikuboyama Volcano (17,000 years ago) and Kawagodaira Volcano (3,200 years ago) in the city of Izu.

Next, I will focus on the types of magma ejected from each volcano. From 150,000 to 20,000 years ago, volcanoes ejected a large amount of relatively low-viscosity basalt and andesite. However, only from 20,000 years ago (virtually, from 3,200 years ago when Kawagodaira Volcano erupted) did high-viscosity rhyolitic magma begin to erupt.

The period of the Izu Tobu Volcano Group

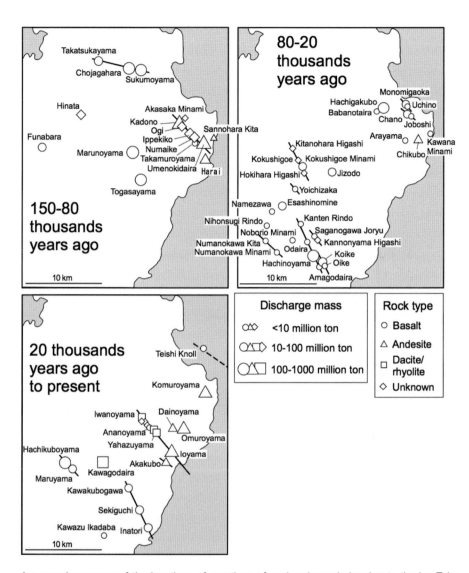

A mapped summary of the locations of eruptions of each volcano belonging to the Izu Tobu Volcano Group, discharge mass of magma and types of discharged magma (rock types). Volcanoes whose eruption period is unclear are excluded.

57. A summary of eruptive history (2): Serious future

In this section, I will focus on the scale of volcanic eruptions. The length of the bold vertical bars in the illustration shows the scale of each eruption (discharge mass of magma) and the marks at the tips of the bars show magma types (rock types). As noted in the previous section, the eruption of rhyolitic magma with square marks can only be seen at the right edge (for the last 3,200 years) of the illustration. It is quite clear that the illustration is filled with round marks (basaltic magma) and triangular marks (andesitic magma).

In the period of time earlier than 100,000 years ago, there were a lot of relatively large eruptions, such as the Kadono-Umenokidaira volcano chain and the Takatsukayama-Sukumoyama volcano chain. In the meantime, the illustration shows that from 100,000 to 40,000 years ago, it was a quiet time with small eruptions. However, large eruptions began again after the eruption of Hachinoyama Volcano 36,000 years ago. In particular, in the last 4,000 years, there occurred a succession of large eruptions, including Omuroyama Volcano (510 million tons), Kawagodaira Volcano (760 million tons) and the Iwanoyama-Ioyama volcano chain (a total of 360 million tons).

These trends can be made further clear by looking at a step diagram drawing the total accumulation of discharge mass of magma since the birth of the Izu Tobu Volcano Group. On this diagram, the period in which there was no eruption draws a horizontal line and the period in which there was an eruption draws a step. The height of steps shows the discharge mass of magma. In the period when average magma discharge rates were low, the steps are gradual and in the period when average magma discharge rates were high, the steps are steep. The steps until 100,000 years ago are steep and the steps from 100,000 to 40,000 years ago are gradual. However, the steps become steep again past 40,000 years ago and the steps have become so sharply steep that they cannot be climbed since 4,000 years ago. This means that the Izu Tobu Volcano Group is currently in a period in which a rapid succession of large eruptions are more likely to occur and that a large eruption may happen at any time from now onward.

The period of the Izu Tobu Volcano Group

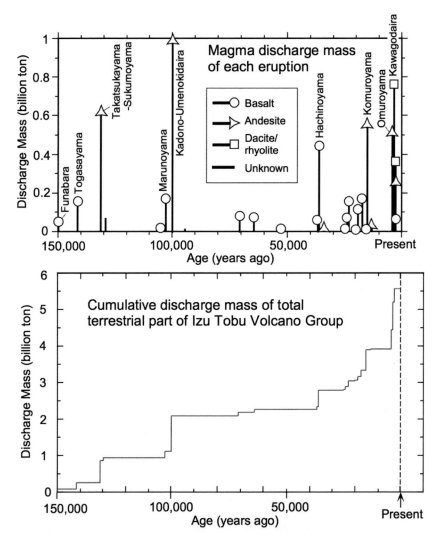

The illustration above shows the eruption age and the discharge mass of magma of each volcano counted among the Izu Tobu Volcano Group. The illustration below shows temporal changes (step diagram) in the total accumulation of discharge mass of magma with a focus on the whole Izu Tobu Volcano Group.

58. A summary of eruptive history (3): Doughnut-shaped structure

Where and how does the magma reservoir of the Izu Tobu Volcano Group exist? The type of magma discharged from each volcano is the key to finding the answer to this question. On the outer side of the distribution of the Izu Tobu Volcano Group exists the basaltic zone in which only low-viscosity basaltic magma is discharged. Inside of the basaltic zone is the andesitic and rhyolitic zone in which high-viscosity andesitic and rhyolitic manga is discharged in addition to basaltic magma. Andesitic and rhyolitic manga is created when basaltic magma melts part of the crust or different types of magma mixed with each other. In addition, studies of crustal structure by seismic waves led to the speculation that there is an accumulation of magma in a broad range of the eastern part of the Izu Peninsula 15 kilometers underground.

Based on these data, it is considered that the doughnut-shaped structures formed by the basaltic zone and the andesitic/rhyolitic zone probably show the distribution itself of underground magma. Conceivably, the reason why andesitic and rhyolitic magma exists only in the inner zone is that basaltic magma that has gradually risen since the birth of the Izu Tobu Volcano Group melted a larger amount of crust in the inner zone with a greater thermal value. However, each single magma reservoir is small and its combinations have not progressed. Therefore, as observed with the eruption of the Iwanoyama-Ioyama volcano chain 2,700 years ago, different types of magma can sometimes erupt on the same eruptive fissure.

What I mentioned above is very significant in terms of disaster damage control as well. Because high-viscosity magma often causes explosive eruptions, you should be more vigilant against eruptions in the andesitic and rhyolitic zone than against eruptions in the basaltic zone. It was fortunate that the eruption of the Teishi Knoll in July 1989 occurred in the basaltic zone, but there is no guarantee that magma activities will still remain off the coast of Ito in the future as well. It is necessary to constantly note where earthquake swarms indicating underground magma activities will occur and whether the location will be within the andesitic and rhyolitic zone or not.

The period of the Izu Tobu Volcano Group

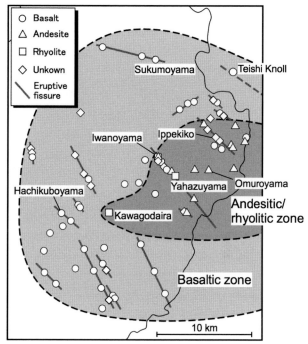

Types of magma discharged from each volcano of the Izu Tobu Volcano Group and doughnut-shaped structure

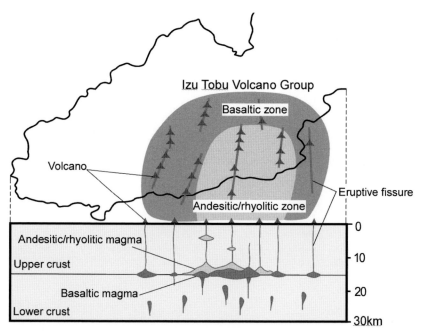

The crustal structure of the Izu Tobu Volcano Group

Chapter 5

The living earth of Izu (earthquakes and crustal movement)

59. The current situaiton surrounding Izu in terms of earth science

With the exception of the eruption of the Teishi Knoll on the sea floor off the coast of Ito in July 1989, volcanic eruptions in Izu and its neighboring waters have not been known since the eruption of the Iwanoyama-Ioyama volcano chain approximately 2,700 years ago. However, the earth of Izu still continues to be active and there have been frequent occurrences of large earthquakes. Some of those earthquakes caused cracks and displacements in the earth and in some cases, clear landforms of active fault can be observed as a result of multiple displacements.

As noted in Section 1 of this book, the earth of Izu is located in an extraordinary place in terms of earth science. Four plates lie in a pile around the Japanese archipelago and Izu sits on the northern tip of the Philippine Sea Plate. The Philippine Sea Plate on which Izu sits is slowly drifting northwestward toward the Japanese main island at a few centimeters per year. The Philippine Sea Plate is subducting under the plate on the side of the Japanese main island on both sides of the Izu Peninsula and the abrasion between plates causes occasional mega-earthquakes. The interplate earthquake occurs repeatedly along the Suruga Trough is called the Tokai Earthquake. The interplate earthquake occurs repeatedly along the Sagami Trough is called the Kanto Earthquake. In addition, there is a fracture inside of the Philippine Sea Plate off the northeastern Izu Peninsula and it is the region where the Odawara Earthquake (the West Kanagawa Prefecture Earthquake) occurs. These major earthquakes violently shook the earth of Izu and caused tsunamis to reach the coast. Those areas suffered serious damage every time they were struck by large earthquakes.

Meanwhile, because the crust around Izu has been warmed by volcanic heat and has become buoyant, it cannot easily subduct under other plates. This is why Izu, which used to be a floating island in the South Seas, is now peninsula-shaped after having collided with the Japanese main island. However, because force is working to push Izu into the Japanese main island, many active faults have arisen on the Izu Peninsula and in its neighboring areas and their occasional activities cause major earthquakes. For example, the Tanna Fault stretching from Kannami Town to the city of Izunokuni and the Irozaki Fault in Minami-Izu Town are famous active faults.

The living earth of Izu (earthquakes and crustal movement)

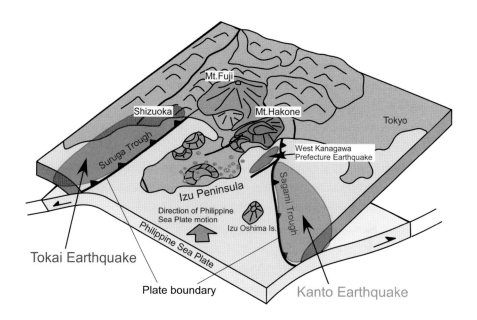

The current situation surrounding the Izu Peninsula and its neighboring areas in terms of earth science. The locations of major earthquakes in neighboring waters are also shown.

Rock monument in the temple Todenji in the city of Shimoda. This monument was constructed for the tsunami victims caused by the 1854 Ansei Tokai Earthquake.

60. The Tokai Earthquake, the Nankai Earthquake and the Kanto Earthquake

With regard to earthquakes that caused damage to Izu, first of all, I will explain the interplate Tokai, Nankai and Kanto Earthquakes. The Tokai Earthquake and the Nankai Earthquake are mega-earthquakes that occur along the plate boundaries on the western side of Izu. The hypocentral region of the Tokai Earthquake ranges from the Suruga Bay to the Kumano Nada (E, D and C on the map). The hypocentral region of the Nankai Earthquake ranges from the Kii Suido Channel to Off Shikoku (B and A on the map). In this connection, in recent years, an earthquake whose hypocentral region ranges from the Suruga Bay to Off Omaezaki (E on the map) has been called the Tokai Earthquake and an earthquake whose hypocentral region ranges from the Enshu Nada to the Kumano Nada (D and C on the map) has been called the Tonankai Earthquake. However, these are administrative names and are not based on the whole earthquake history. Even if an earthquake whose hypocentral region is a single place (A to E) on the map occurs, it will become a magnitude of nearly 8 on the Richter scale. If multiple earthquakes simultaneously occur at all the hypocentral regions (A to E), they will be ultra-mega-earthquakes measured 8.7.

As is shown in the map, the hypocentral regions of the Tokai Earthquake and the Nankai Earthquake are located next to each other and the occurrence patterns are also closely linked. A detailed examination of old records has revealed that these two earthquakes occurred simultaneously or at an interval of two years at longest. To put it another way, the Tokai Earthquake and the Nankai Earthquake are twin earthquakes and are apt to occur almost at the same time in succession. The interval of occurrence based on the assumption that the two earthquakes are twins is from 90 to 200 years. Records of their occurrences even go back to the Asuka period (684), but it was after the Meio Tokai-Nankai Earthquake of 1498 that there appeared clear records on earthquake disasters in Izu whose historical materials are quite scarce.

In the meantime, the Kanto Earthquake has repeatedly occurred along the plate boundary (F on the map) on the eastern side of Izu. For historical records on the Kanto Earthquake, the Taisho Kanto Earthquake of 1923, which caused the Great Kanto Earthquake disaster, and the Genroku Kanto Earthquake of 1703 in the Edo period are well-known. There remain many records of serious damage caused by these two earthquakes due to their strong shakes and tsunamis. Unfortunately, for the occurrence history of the Kanto Earthquake before medieval times, there are only

The living earth of Izu (earthquakes and crustal movement)

scarce historical materials on the Kanto area and many things are unclear, including the interval of occurrence. Earthquake records on 878, 1293 and 1495 are counted among the candidates for the Kanto Earthquake, but it is considered that there are still buried records.

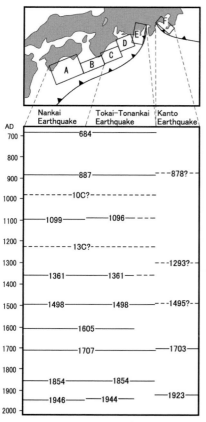

Mega-earthquakes that occur on plate boundaries along both sides of Izu. The triangular bold lines show plate boundaries and the square A to F show hypocentral regions. The chronological figure below shows the years in which each earthquake occurred.

A stone signpost in the precincts of Kaizoji Temple in the city of Ito, which shows the height of the tsunami caused by the Taisho Kanto Earthquake of 1923

61. The West Kanagawa Prefecture Earthquake

In the previous section, I discussed the Tokai Earthquake, the Nankai Earthquake and the Kanto Earthquake whose hypocentral regions are located along the plate boundaries stretching on both the eastern and western sides of Izu. Although different from those interplate earthquakes, the West Kanagawa Prefecture Earthquake (the so-called Odawara Earthquake) has important characteristics. A fault (West Sagami Bay Fracture) that is considered the source of this earthquake is located a few or dozens of kilometers underground from the waters off the northeastern coast of the Izu Peninsula and is also considered to be a fracture inside of the Philippine Sea Plate. The Philippine Sea Plate is subducting under the Kanto district from the Sagami Trough, while Izu cannot subduct because of its collision with the Japanese main island. In this situation, it is the West Sagami Bay Fracture that was created by the separation between the subducting Philippine Sea Plate and Izu. The West Sagami Bay Fracture is, so to speak, a fissure that has begun to cut the bedrock off the eastern coast of Izu from the north as if with scissors. For the future, the West Sagami Bay Fracture is considered to spread south-southwestward between the Izu Peninsula and the Izu Shichito Islands and to eventually grow into a plate boundary fault that cuts the whole Izu Peninsula.

As noted above, the hypocentral region of the West Kanagawa Prefecture Earthquake is situated just next to that of the Kanto Earthquake and it is conceivable that these two earthquakes occurred in complicated association with each other. The oldest earthquake that can be counted among the West Kanagawa Prefecture Earthquake is the Kan'ei Odawara Earthquake of 1633. This earthquake was at a level of 7 on the Richter scale. It was triggered by the rupture of the southern side of the West Sagami Bay Fracture and caused a tsunami damaging the coast from the city of Atami to the city of Ito. The Tenmei Odawara Earthquake of 1782 and the Kaei Odawara Earthquake of 1853 can also be considered to have occurred on the West Sagami Bay Fracture just like the Kan'ei Odawara Earthquake of 1633. Because the northern side of the West Sagami Bay Fracture ruptured, these two earthquakes seem not to have caused a clear tsunami.

Meanwhile, some think that the West Sagami Bay Fracture was also active at the same time as when the Genroku Kanto Earthquake of 1703 and the Taisho Kanto Earthquake of 1923, mentioned in the previous section, occurred. Unless you think that way, you cannot explain the characteristics of the tsunami and the uplifts of the

The living earth of Izu (earthquakes and crustal movement)

Hatsushima Island and Cape Manazurumisaki at the occurrence of the earthquakes. That is, strictly speaking, these two earthquakes were simultaneous occurrences of the Kanto Earthquake and the West Kanagawa Prefecture Earthquake. In accordance with this line of reasoning, earthquakes that occurred on the West Sagami Bay Fracture can be counted as five times in the past, 1633, 1703, 1782, 1853 and 1923, which means that earthquakes occurred at the average interval of 73 years. Based on this estimate, the next West Kanagawa Prefecture Earthquake was projected to occur around 1998, but fortunately, we have yet to see such earthquake occur.

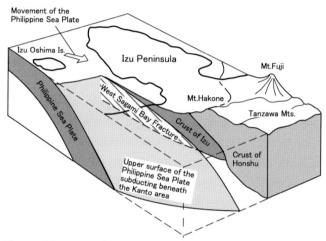

The crustal structure in a range from Izu to the Sagami Bay and the West Sagami Bay Fracture. Based on the original illustration drawn up by Honorary Professor Katsuhiko Ishibashi of Kobe University.

A bird's-eye view of the city of Odawara and its neighboring areas. The West Sagami Bay Fracture is considered to be located deep underground around this area.

62. The Tanna Fault (1): Disconnected ravines

An active fault is a slip of an earthquake source fault deep underground that reached the earth surface and can be clearly recognized as displacements of landforms and strata. Therefore, underground an active fault lies an earthquake source fault capable of potentially causing a major earthquake. It is known that there are many active faults on the Izu Peninsula and in its neighboring waters.

Probably, the Tanna Fault is the most famous active fault in Izu. Its landform is quite clear. Starting from the southern side of the Hakonetoge Pass, the fault stretches 18 kilometers from the north to the south, running through the Tanna Basin in Kannami Town to around the Ukihashi area in the city of Izunokuni. In addition, a fault group that can be considered an extension of the Tanna Fault stretches 13 kilometers southwestward from around the Ukihashi area to around Lake Sagiriko in the city of Izu (see Frontispieces 9 [above] and [below]).

The first person that insightfully discovered the astonishing secret behind the Tanna Fault was Hisashi Kuno (who later became a professor at the University of Tokyo), whom I noted in Section 31. He focused on the landform displacement between three particular ravines. All three ravines stopped short just at the Tanna Fault and was disconnected ahead of their eastern side. If the Tanna Fault were just vertically displaced, the three ravines should be connected with their eastern side, but things are actually quite different. Kuno speculated about imaginary ravines (A' to C') that should have connected with the three ravines (A to C) from their landforms. If you assume that the Tanna Fault vertically slid the land on both of its sides one kilometer, the three ravines would be perfectly connected. In addition, Kuno discovered that the boundary between Yugawara Volcano and Taga Volcano was also dislocated by the Tanna Fault to the same degree.

Based on these sets of evidence, Kuno claimed that the Tanna Fault had a large lateral dislocation of at least one kilometer in addition to a vertical dislocation of approximately 100 meters. Surprisingly, he presented this analysis in his paper published in 1935. At that time, the whole world did not know that there was a fault with such a large lateral dislocation. In addition, it had not been made clear yet that fault dislocation caused earthquakes. In this sense, Kuno's research results were extremely progressive and pioneering.

The living earth of Izu (earthquakes and crustal movement)

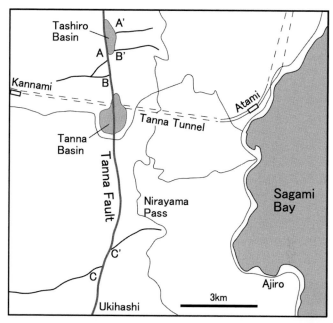

The landform displacement of ravines along the Tanna Fault

A 3-D model of the northern landform of the Tanna Fault in the Tanna Fault Park, the Kannami Town. The arrows of opposite directions show the displacement of the fault.

63. The Tanna Fault (2):
The Kita Izu Earthquake and trench excavation

A large earthquake struck northern Izu before daybreak on November 26, 1930. This is the Kita Izu Earthquake of 1930 (a magnitude of 7.3 on the Richter scale). This earthquake caused lateral dislocations to the Tanna Fault, several faults to the southwest of it and the Himenoyu Fault (Himenoyu in the city of Izu) with some places laterally dislocated by more than two meters.

This lateral dislocation also directly hit the construction site of the Tanna tunnel on the Tokaido Line, which had been dug underground the Tanna Basin at that time. The tunnel was dug from both eastern (Atami) and western (Kannami) sides. When construction workers had dug the tunnel on the western side up to the Tanna Fault, they suspended their operation because of stronger water springs and were digging water-removing pipes along the main tunnel. Three of those pipes were displaced by two meters by the dislocation of the fault and the front part of the tunnel was completely blocked by the slid bedrock.

Afterward, the tunnel construction continued and this tunnel is now part of Japan's main artery connecting between the east and west of the nation together with the Shin Tanna Tunnel for bullet trains, which was later built as an annex. However, probably, everyone wants to know how distant the future will be in which the Tanna Fault's next dislocation causes the tunnel to be dislocated. In addition, the Kita Izu Earthquake registered more than 6 on the Japanese seismic scale in a broad range of areas in the northern part of the Izu Peninsula, even registering 7 in some places. The earthquake also completely destroyed more than 2,000 houses in Shizuoka prefecture alone and took a death toll of nearly 250. Probably, every citizen in Izu is interested in the interval of repeated occurrences of such large earthquakes.

It was in the early 1980s, 50 years after the Kita Izu Earthquake, that a trench excavation of the Tanna Fault was carried out to know the interval of repeated occurrences of earthquakes. A trench excavation of faults is usually conducted selectively in a place where sand and mud often pile up, such as basin and depression. When land gets stepped under the influence of a fault dislocation, the steps are gradually filled in by sand and mud and the land eventually reverts to its original flat form. During that period of time, sand and mud are more likely to thickly pile up on lower steps, which results in creating strata with different thickness on both sides of the steps. To put it another way, immediately after a fault

The living earth of Izu (earthquakes and crustal movement)

dislocation, strata are often different in thickness on both sides across from the fault. Therefore, it is possible to broadly estimate the period of a fault dislocation by digging out such proper strata and examining their geologic age.

A fault dislocation caused by the Kita Izu Earthquake can be seen even today. The Tanna Fault runs through underground of the dashed-line part. The stone fences marked A and A' and the waterways marked B and B' are dislocated by one meter in the same direction. The semicircular stonework marked CD and C'D' was originally a circle. The picture was taken at the Tanna Fault Park in Kannami Town.

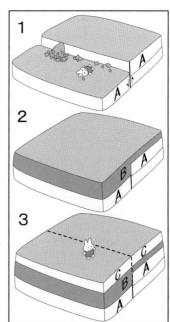

How strata pile up around an active fault. (1) The slide of the active fault dislocates the A layer and creates a step on the ground. (2) The B layer different in thickness piles up on both sides across from the step. (3) The C layer piles up after the step is filled in. After a while, the active fault slides again. Another step is formed at the dashed-line part. The C layer is dislocated and the strata revert to (1).

64. The Tanna Fault (3): The past and the future

As its name shows, a trench excavation of an active fault is carried out by directly digging a certain range of ground including a fault by excavation machine. At an excavation site on the northern edge of the Tanna Basin, strata six meters deep from the surface of the ground were dug out. As expected, displacements by the Tanna Fault's dislocation were actually observed there. Evidence of a total of nine large earthquakes was obtained and the periods of their occurrences could be calculated by the method I explained in the previous section.

First of all, needless to say, the Kita Izu Earthquake of 1930 was the most recent earthquake of the nine ones. For the second most recent earthquake, evidence was later discovered in a trench excavation of the Tashiro Basin to the north of the Tanna Fault and it became clear that the earthquake occurred during the period between the late 13th and early 17th centuries. This period corresponds to medieval times about which there are just scarce historical materials of Izu.

The occurrence of the third most recent earthquake could be identified as 841. It was discovered that excavated strata contained a layer of volcanic ash that piled up by the huge eruption of Kozushima Volcano on the Izu Shichito Islands in 838 and it emerged that a large-scale earthquake occurred immediately after that eruption. Shoku Nihon Koki, an official chronicle compiled by the Japanese Imperial Court in the ninth century, includes descriptions of this eruption of Kozushima Volcano. Subsequently, some descriptions of the damage Izu suffered from an earthquake were found and it was discovered that a large earthquake occurred around the spring of 841. Although there were no descriptions of specific damaged places, there remained detailed accounts of relief efforts. Based on these historical descriptions, it is speculated that the central area of Izu at that time, that is, the northern Izu region, was particularly damaged and it seems logical to think that the huge earthquake was caused by the Tanna Fault.

In this way, it was discovered that there had been a total of nine earthquakes for the last 8,000 years and it became clear that the Tanna Fault caused large earthquakes at the average interval of 1,000 years. With a focus on the three most recent earthquakes whose frequency of occurrence is relatively high, the average interval is 540 years. It has been just 80 years or so since the last Kita Izu Earthquake of 1930 occurred. Therefore, it may be possible to say that the Tanna Fault will not cause a huge earthquake for the next 500 years to come.

The living earth of Izu (earthquakes and crustal movement)

A trench excavation of the Tanna Fault carried out in the middle part of the Tanna Basin in 1985.

The Tanna Fault dug out by the trench excavation. Dislocated strata can be clearly observed between the right and left sides of the fault.

65. A country of active faults

It is known that there are many active faults, including the Tanna Fault mentioned in the previous section, on the Izu Peninsula. For active faults whose activity in historically large earthquakes has been confirmed, they include the Tanna Fault for the Kita Izu Earthquake of 1930, the Irozaki Fault for the Izu-hanto-oki Earthquake of 1974 and three particular faults around Inatori for the Izu-Oshima-kinkai Earthquake of 1978. However, the primary fault that caused the Izu-Oshima-kinkai Earthquake is located on the sea floor between the Izu Peninsula and the Izu Oshima Island and the three particular faults around Inatori are just subordinate ones. For other active faults, there are just scarce historical records of Izu before the Meiji period (1868-1911) and the history of their activity has hardly been clarified.

The Kamigamo Fault runs parallel to the Irozaki Fault to the north of it, stretching from around the Jaishi area in Minami-Izu Town to around the Toji area in the city of Shimoda. Although historical records of earthquakes in this area are extremely limited, it is known that an earthquake that occurred on March 8, 1729, caused damage to the city of Shimoda and Minami-Izu Town. The intensity of this earthquake was similar to that of the Izu-hanto-oki Earthquake of 1974 and it can be considered to have registered a magnitude of 7 on the Richter scale. In addition, because the activity interval of the Irozaki Fault is considered to be approximately 1,000 years, it is difficult to ascribe the occurrence of the earthquake in 1729 to the Irozaki Fault. Therefore, it can be suspected that the Kamigamo Fault might have caused the earthquake in 1729.

Around the Kadono area in Matsuzaki Town is the Kadono Fault that stretches in northeastern and southwestern directions. This fault has hardly been paid attention to as an active fault. However, geologically, it is a significant fault and there remain landforms of large landslides that seem to have collapsed by earthquakes around the fault. It seems to be an imminent challenge to find the truths behind the history of the fault activity, including how old the remaining landforms are.

The Mizunuki-Yoichizaka Fault stretches in northwestern and southeastern directions from Mizunuki to Yoichizaka around the Yugashima area in the city of Izu. Geologically, this fault is also clear and significant. However, its activity history is totally unclear and it is necessary to pay careful attention to the Mizunuki-Yoichizaka Fault just like the Kadono Fault.

The living earth of Izu (earthquakes and crustal movement)

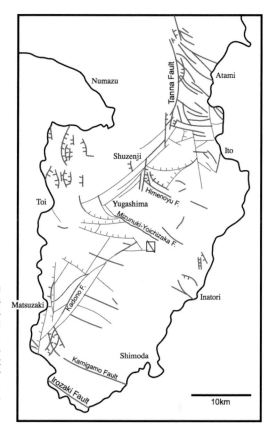

Primary active faults on the Izu Peninsula. The bold lines show faults recognized as active faults. (The ones marked by fluffs are normal faults and the other ones are laterally dislocated faults.) Evidence of active faults has yet to be obtained, but main faults that have been identified through my geologic surveys are shown by thin lines. Small rectangle shows the range of the photo below.

Hatchoike Pond near Mt.Amagisan. The pond filled a depression, which is made by the movement of an active fault (white dashed line).

66. The mystery of tectonic rotation: Unusual fault distribution

As clearly shown in the illustration of active faults on the Izu Peninsula, mentioned in the previous section, the crust on the eastern side of the Tanna Fault, which is a major active fault on the peninsula, is finely chopped like stripes of paper by many other active faults that stretch in northwestern and southeastern directions. The stripes of paper are even more than ten kilometers long on land alone, but they are just hundreds of meters or one kilometer in width. The distribution of such remarkably unique active faults is quite rare in the whole of Japan and it can be said to form an unusual crustal structure. What in the world happened to the crust in the northeastern part of the Izu Peninsula?

The key to finding the hidden truths behind this crustal structure was the weak paleomagnetic azimuth of rocks. As explained in Section 25, by measuring this azimuth, I discovered that when Izu collided with the Japanese main island and indented into it, Izu rotated its neighboring areas, such as Oiso Hill in Kanagawa, from their roots. I also said in Section 25 that for Izu, there were no large tectonic rotations at least after five million years ago with the exception of some areas. However, "some exceptional areas" refer to the area on the eastern side of the Tanna Fault, which I focus on in this section.

Through our measurement in the 1980s, we discovered that the whole area sandwiched between the Tanna Fault and the east coast of Izu was rotating in a more or less counterclockwise direction. The rotation of the area along the coast from Ajiro in the city of Atami to Usami in the city of Ito and its inland is particularly notable. Some areas show a rotation angle of more than 70 degrees.

Many of the strata distributed in this area are lava flows effused from Taga Volcano and Usami Volcano, mentioned in Section 31. The lavas effused from volcanoes chill and harden in a short period of time and magnetic minerals, such as magnetite contained in the lavas, make records of the direction and intensity of a geomagnetic field. Therefore, normally, the lava flows with the same geologic age show a consistent magnetic direction whatever spot of them is measured. In some cases, the direction and intensity of magnetism change later. However, because we extracted only stable components by laboratory treatments, it is difficult to consider this as the cause of the deviation of magnetic directions. Although the deviation of magnetic directions can also be caused by the tilts of strata, many of the lava flows in this area

The living earth of Izu (earthquakes and crustal movement)

are almost horizontal or gradually sloping. That is why I have to consider that a major reason for the deviation of magnetic directions is that the bedrock itself, including the sampling points, rotated.

What caused this rotation of the bedrock? Why does only the eastern part of the Tanna Fault rotate?

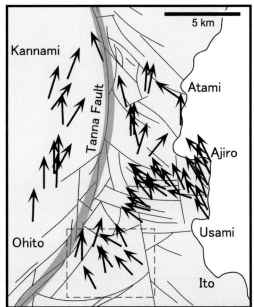

The arrow marks show the magnetic directions of lavas at sampling points of Taga Volcano and Usami Volcano. These arrows should align to the direction of geomagnetic field, that is, directions near the north, but many arrows on the eastern side of the Tanna Fault (the gray bold line) show counterclockwise directions. The thin lines show main active faults.

The author in his university days is collecting rock samples of Usami Volcano by a converted drill from a chain saw.

67. The mystery of tectonic rotation: The mechanism of rotation

The evidence that the crust of the area sandwiched between the Tanna Fault and the coastline to the east of it had undergone a tectonic rotation was obtained from the magnetic measurement results of the lavas effused from Taga Volcano and Usami Volcano. It can be considered that the rotation occurred during the period of time from approximately 500,000 years ago after the lava effusion to the present.

The crust of the area mentioned above took a long time to gradually rotate, but surprisingly, it was only in the 1980s that such a large-scale movement of the earth became known. As I explained in Section 25, if the earth gets inclined or bends, it can easily be recognized as dips of strata with the naked eye. However, if such crustal deformation do not involve dips of strata, they are difficult to detect. For another example of the detection of large-scale tectonic rotation through rock magnetism measurement, there is a well-known preceding study proving that the Japanese archipelago bent into its current shape along with the expansion of the Sea of Japan 15 million years ago.

Compared with such large-scale rotations, the rotation of the eastern side of the Tanna Fault is limited to a particular area. What caused this rotation? Given the fact that the rotation is limited to the eastern side of the Tanna Fault, the activity of the fault can be suspected to have caused the rotation. After having examined similar studies carried out all over the world, I knew that foreign researchers theorized the mechanism in which an area sandwiched between two laterally dislocated faults gets many fissures by the force of distortion generated by movements in opposite directions and the crust chopped into shapes like strips of paper by the fissures gradually rotates. I realized that groups of faults shaped like strips of paper in the same way as Izu had been discovered in many places all around the world with the detection of large-scale tectonic rotations by rock magnetism measurement and in-depth examinations of their causes.

If the hypothesis of this mechanism is correct, another pair to the Tanna Fault must lie somewhere on the seafloor off the coast of Atami. It is the West Sagami Bay Fracture noted in Section 61 that can be suspected as a significant candidate. In addition, it can be assumed that the rotation of the crust shaped like strips of paper may create triangular chinks (the gray parts in the illustration) on the eastern side of the Tanna Fault. Probably, a line of basins close to the east of the Tanna Fault, such

The living earth of Izu (earthquakes and crustal movement)

as Tanna, Tashiro, Tawarano and Ukihashi Basins, were created by earth depressions triggered by the formation of such openings (see Frontispiece 9 [below]).

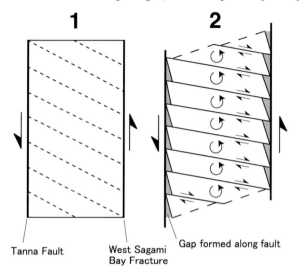

The mechanism in which the eastern crust of the Tanna Fault rotates. (1) The crust of the area sandwiched between the Tanna Fault and the West Sagami Bay Fracture gets many echelon fissures (the dashed lines in the illustration) under the influence of a force that works to dislocate the fault. (2) As things progress, all the echelon fissures are dislocated in the form of faults and the crust chopped into shapes like strips of paper rotates.

A bird's-eye view of the Tanna Fault and a line of basins to the east of its southern extension. The picture was taken by the Izu Peninsula Geopark Promotion Council.

68. Coastal landforms tell

Although tectonic rotations and the movement of micro-plates are dynamic, their progress takes a long time and is difficult to recognize by the naked eye. However, the historical movement of the earth is sometimes recorded in our familiar landscapes.

When we walk along the water's edge on the beach, we sometimes see uniquely shaped landforms—wave-cut bench and wave-cut notch. As their names show, a bench and a notch mean a flat plane and a dent created by wave erosion. A cliff on the beach gradually recedes to the side of land when it is cut by waves and benches form after the cliff recedes. At the bottom of a receding cliff are notches washed by waves. You can find benches and notches on the water's edge on a rocky beach at low tide. Benches are hidden under the surface of the water at high water and waves roll in to where notches are.

However, after I had closely examined the east coast of Izu, I found that I could see benches and notches in high places where waves do not reach even at high tide. This means that these wave-cut landforms were created in a period when the surface of the sea was higher than today or the land itself was uplifted after the creation of the landforms. It is known that the surface of the sea was three meters higher in the Jomon period approximately 7,000 years ago than today due to global warming. Therefore, there is some possibility that wave-cut landforms created during that period of time remained in high places due to the decline of the surface of the sea in later years. To prove this, it is necessary to know when the wave-cut landforms were created.

A wave-cut landform itself is just the bumpiness of the rock surface and it is difficult to directly examine its geologic age. However, creatures with calcareous shells, such as shellfish, barnacle and lugworm, sometimes stick to the dents of rocks. The geologic age of these shells can be measured by the radiometric carbon dating method mentioned in Section 3.

The studies that have been made so far demonstrated that fossils sticking to the notches on the east coast of Izu were younger than 7,000 years old and also observed two layers of fossils that were different in altitude and geologic age. This means that the east coast of Izu underwent phased uplifts by earthquakes twice after 7,000 years ago. However, the scale of those earthquakes, the locations of faults that caused the earthquakes and the interval of repeated occurrences have yet to be clarified. In

The living earth of Izu (earthquakes and crustal movement)

terms of disaster damage control, it is appropriate to assume that unknown active faults lie somewhere on the sea floor in the waters from Shimoda to Ito.

An example of wave-cut landforms. The flat plane on which four persons are standing is a bench. The dented part just behind the two persons on your right is a notch. The picture was taken around Cape Shiofukizaki in the city of Ito. The movement of the earth is hidden in such an ordinary coastal landform as this.

Calcareous fossils sticking to a sea cave on the coast in the city of Ito. They prove that the coast was uplifted.

69. A peninsula inclined to the west

A careful visitor to southern Izu would realize that there are strange differences in landform between the eastern and western sides of the peninsula. For example, if you drive along a prefectural road that runs from Shimoda through the Basaratoge Pass to Matsuzaki on the west coast, you will see a stretch of flat land 15 kilometers long across which the Inozawagawa River quietly flows until you come to the Kazono area in the city of Shimoda. However, once you travel through the Basara Tunnel, you will see a dramatic change in scenery. You drive approximately five kilometers along a steep mountainous path and then, you will see the road approximately five kilometers long to the Matsuzaki Coast running in a flat land across which the Nakagawa River flows.

You can see a more extreme change in landform if you drive from the Yumigahama area in Minami-Izu Town through the Shimogamo area to the Mera area on the west coast (see Frontispiece 10 [below]). The ten-kilometer section from Yumigahama to Tateiwa, which constitutes 90% of the whole route, is a flat land through which the Aonogawa River and its branch streams gently run. In contrast, once you go through the Mera Tunnel, you go down a sharp slope 1.5 kilometers long until you come to the port of Mera.

That is, the landform in southern Izu is smooth on its eastern side and it is distant from the coastline. In contrast, the landform in southern Izu is sharp on its western side and it is close to the coastline. These asymmetrical landforms between the eastern and western sides are sometimes created by different geologic characteristics and rocks. However, the geologic nature in southern Izu is almost homogeneous (many of the strata belong to the Shirahama Group explained in Sections 12 to 18) and it is necessary to elaborate a different explanation.

A significant clue to approach this question is the uplift of the east coast of Izu indicated by coastal landforms, as explained in the previous section. If Izu's east coast had gradually been uplifted on one hand and the west coast had sunk on the other, what would have happened to their landforms? The plains on the west coast would have sunk into the Suruga Bay and the slope on the western side of mountains would have become sharper. In contrast, the plains on the east coast would have expanded to the side of the Sagami Bay and the slope on the eastern side of mountains would have become more gradual.

I obtained several data that support this speculation. There used to be lagoons

The living earth of Izu (earthquakes and crustal movement)

along rivers pouring into the east coast, such as the Inozawagawa River, the Aonogawa River and the Ito Okawa River. The highest point of strata that piled up in those lagoons was higher than the current surface of the sea. In contrast, the altitude of the same strata around Matsuzaki was lower than the current surface of the sea. In addition, you can observe some flat planes (coastal terrace), which are the remnants of old coastal plains, along the coast on the Izu Peninsula. For the height of flat planes formed in the same period, they are high on the side of the Sagami Bay and low on the side of the Suruga Bay (see Frontispiece 10 [above]).

Based on these facts, it is considered that the east coast of Izu is being uplifted and the west coast is sinking. That is, the Izu Peninsula is gradually dipping westward. Probably, this dip is caused by plate subduction along the Suruga Trough. Izu is being slowly dragged into the depth of the Suruga Trough along with plate motion.

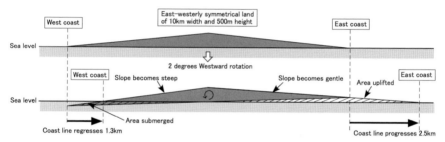

How the landform in southern Izu was created. Imagine there is land with a width of ten kilometers and a height of 500 meters. The illustration simulates a change of coastline and slope supposed that the land tilts to the west by two degrees.

A bird's-eye view of the asymmetrical landform of southern Izu. It is distant from the east coast to the Meratoge Pass, the watershed.

Chapter 6

The living earth of Izu (magmatism)

70. The genealogy of volcano deities

The Shirahama Shrine in the city of Shimoda is located on the Shirahama Coast, which I introduced as the "Pliocene Coast" in Izu in Section 12. This shrine is devoted to the Ikonahime Deity. Nihon Koki, an official chronicle compiled by the Japanese Imperial Court in the ninth century, included a description of a mysterious episode about this goddess in the Heian period (794-1185). Its original text was lost in a long history, but fortunately, another book cited the story and we can read it today. According to the book, in 832 the Ikonahime Deity closed a deep ravine, broke a tall rock and created approximately 2,400 hectares of flat land, two "ins" (probably, small hills) and three ponds. At that time, the Imperial Court sought to soothe the rage of this violent goddess and designated the Ikonahime Deity and the Mishima Deity as special deities to be enshrined. The Mishima Deity was the husband of the Ikonahime Deity and is enshrined at the Mishima Shrine in the city of Mishima.

The above-quoted description "closed a deep ravine" must have meant a volcanic eruption. This is because when a volcano erupts, lavas effused from the crater often flow burying a ravine and create a lava fan on the coast. In addition, small volcanoes (such as scoria cone I repeatedly mention in this book) are formed around the crater and if water accumulates in the crater, it results in creating a lake and a pond. If a fissure eruption occurs, it creates a line of small volcanoes and craters.

Did a volcanic eruption occur on the Izu Peninsula in 832? If so, it means that the Off Ito Submarine Eruption of 1989 was not the first eruption in Izu after historic times. However, no new volcanoes have been discovered around the Shirahama Shrine and it is considered that the last terrestrial volcanic eruption of the Izu Tobu Volcano Group was that of the Iwanoyama-Ioyama volcano chain approximately 2,700 years ago, which I explained in Section 55.

You can find a key clue to uncover the truth behind the 832 incident in the history of the Shirahama Shrine. According to historical studies, the Ikonahime Deity used to be enshrined on the Miyakejima Island together with the Mishima Deity. Subsequently, the two deities were enshrined at the Shirahama Shrine and then, only the Mishima Deity was enshrined at the Mishima Shrine. That is, the shrines to enshrine the deities changed. In addition, given volcanological data, the eruption in 832 is interpreted as what happened on the Miyakejima Island (fissure eruption on the northern slope), where the two deities were enshrined at that time.

The living earth of Izu (magmatism)

The story does not end here. Legend has it that the huge eruption of Kozushima Volcano in 838 was caused by the Awa Deity (primary wife of the Mishima Deity), who got angry with the Imperial Court's designation of the Mishima Deity and the Ikonahime Deity as special deities to be enshrined. The Imperial Court knew this by fortune-telling and rushed to designate the Awa Deity as a special deity to be enshrined as well.

The Shirahama Shrine in the city of Shimoda

Tenjosan lava dome created by the eruption of Kozushima Volcano in 838

71. The illusion of eruption

In 1854, Admiral Putyatin of the Russian Imperial Navy and his aides arrived at the port of Shimoda aboard the frigate Diana on a diplomatic mission to negotiate about a treaty with Japan (see Frontispiece 10 [above]). On December 23, 1854, the day immediately after the first negotiation, the Ansei Tokai Earthquake (refer to Section 60) occurred and the Diana was destroyed by a tsunami that struck the port of Shimoda and was unable to cruise on its own. Five days afterward, Putyatin sent Japanese government officials the following message: "Fire rose from a mountain in Izu the night before last. There is no longer any possibility of earthquake and tsunami. You do not need to worry, please."

This quote is from the Shimoda Diary, which was written by Controller Toshiakira Kawaji, a Tokugawa shogunate official who negotiated with Putyatin. In the West at that time, it had not been discovered that the cause of earthquakes was an underground fault dislocation and Westerners vaguely believed that earthquakes were caused by "underground fire (sulfur)." It seems that Putyatin thought that because fire, the cause of earthquakes, had escaped, there would be no more earthquakes and tsunamis.

Then, what were the true colors of this "fire"? The simplest interpretation is a volcanic eruption. However, as mentioned in the previous section, it is not known that there is a new volcano around Shimoda. In addition, Norimasa Muragaki, a Tokugawa shogunate official who stayed at Shimoda, left the place two days after the large earthquake to report it to the government. Muragaki passed the distribution zone of the Izu Tobu Volcano Group, but did not make any record of unusual phenomena related to volcanic eruption with the exception of damage caused by the earthquake and its aftershocks. Fundamentally, none of the Japanese government officials seems to have witnessed "fire." These facts show that it is difficult to regard the "fire" Putyatin saw as a volcanic eruption. Even if it was true, it was likely to be a mountain fire, the burning of a field and the like.

In this connection, the *Diana* was struck by a strong wind on its way to Heda in western Izu (see Frontispiece 8 [above]) to get it mended, drifted to the waters off the coast of the city of Fuji and finally capsized. The Russian party who had lost their ship returned to their country aboard the *Heda*, a ship constructed by ship builders in Heda.

The living earth of Izu (magmatism)

A view seen from the port of Shimoda. The mountain on your right is Mt. Nesugatayama and the triangular mountain on your left is Mt. Shimodafuji. Both mountains were volcanic necks (refer to Section 17) older than two million years ago washed by erosion.

The port of Shimoda struck by a tsumani caused by the Ansei Tokai Earthquake of 1854. The black ship at the right edge is the Russian Imperial Navy frigate *Diana*. The illustration is in the possession of the Central Naval Museum.

72. Earthquake swarms in 1930

It is well-known that clear earthquake swarms occurred around Ito from February to May in 1930. It was later called the Ito Earthquake Swarm of 1930. Felt earthquakes began on the night of February 13 and the number of earthquakes rose from late February to March (first phase) and in May (second phase). Measurements revealed that hypocenters had been concentrated off the coast of the port of Ito to Kawana and that the surrounding ground was uplifted by a maximum of 20 centimeters. These characteristics are quite similar to those of earthquake swarms off the east coast of Izu Peninsula, which began to occur in and after 1978. Therefore, the Ito Earthquake Swarm of 1930 is considered to have been caused by the magma of the Izu Tobu Volcano Group pushing into the underground just like current earthquake swarms off the east coast of Izu Peninsula.

However, the Ito Earthquake Swarm of 1930 was overwhelmingly huge in scale. Many of earthquake swarms after 1978 continued within one month and felt earthquakes just amounted to 494 times even during the most frequent period from June to September 1989 (earthquake swarms involving submarine eruptions off the coast of Ito). The frequency of earthquakes with a magnitude (M) of more than 5 on the Richter scale was within twice per earthquake swarm. In contrast, in the case of the Ito Earthquake Swarm of 1930, it continued for more than three months despite its remission and the frequency of felt earthquakes amounted to as many as 4,015 times from February to May. In addition, there occurred more than ten earthquakes with more than M5 (the maximum of M5.9). The Ito Earthquake Swarm of 1930 shows the past occurrence of active magmatism beyond the imagination of people who just experienced earthquake swarms after 1978. In addition, the above-mentioned uplift of 20 centimeters is almost half of the total of uplifts triggered by earthquake swarms off the east coast of Izu Peninsula after 1978.

These differences in the scale of earthquakes and uplifts are considered to be due to the overwhelming amount (200 million cubic meters) of magma that intruded into the underground when the Ito Earthquake Swarm of 1930 occurred. The amount of magma intruded by earthquake swarms after 1978 was just 20 million cubic meters at the maximum per earthquake swarm.

In spite of such large-scale magmatism, it was fortunate that the Ito Earthquake Swarm of 1930 did not go as far as to cause eruptions. However, it affected a different area. The magma that intruded into the underground around Ito pushed the

The living earth of Izu (magmatism)

crust in northeastern Izu to the north and can be considered to have induced the Kita Izu Earthquake of 1930 with M7.3 (refer to Section 63), which occurred on November 26 the same year.

Number	Date	Duration	Situation
1	May 28 - early July?, 1596	1 or 2 months	Unknown
2	Late March - Late May?, 1737	2 months?	Many tourists scared and returned to home
3	December 29, 1816 - January 20, 1817	1 month	Many strong earthquakes but no damage
4	Early May - July, 1868	2-3 months	Many damages of stone walls and banks
5	February 13 - Late May, 1930	3 and half months	Slight damages of roads and buildings
6	November, 1978 - Present	Sporadically for nearly 40 years	Slight damages of roads and buildings

A list of earthquake swarms that occurred in Izu in historic times (candidates included). It is definite that 5 and 6 were earthquake swarms off the coast of Ito. This is also likely to be the case with 3 and 4. However, because many historical records of Izu before the closing days of the Tokugawa regime are missing, I would like the readers to think that the list is defective.

The rock bed like a screen in the right half of the picture is a dike formed as a result of the cooling and solidification of the Izu Tobu Volcano Group's magma that ripped up through the earth 23,000 years ago. Currently, the dike provides a rich source of water as an underground dam. The picture was taken in the Suidoyama area, the city of Ito.

73. From 1930 to 1978

Abnormal uplifts around Ito caused by the Ito Earthquake Swarm of 1930, which continued from February to May, did not cease either after earthquake swarm activity had completely ended or after the Kita Izu Earthquake of November 1930 with a magnitude (M) of 7.3 had occurred and continued until at least early 1933. That is, the magmatism of the Izu Tobu Volcano Group quietly progressed even after the end of earthquake swarms. However, at that time, geodetic surveys were carried out only along the coast and it is unclear where the center of uplifts was. Subsequently, uplifting turned into sinking and the earth of Ito continued to slowly sink until the early 1970s. During this period of time, overall earthquake activity in Izu was also quiet.

After this silence had been broken by the Izu-hanto-oki Earthquake (M 6.9) in May 1974, abnormal uplifts began to be observed again around Ito in early 1975. Magma awoke. This is because the large earthquake might stimulate magma in dormancy. It was made clear that the center of the uplift zone was located in the Hiekawa area in the city of Izu and it came to be called "abnormal uplift around Hiekawa." At some points, the amount of uplifting had reached 15 centimeters by the end of 1977. However, no earthquake swarms occurred around the uplift zone. Instead, there were continued occurrences of sporadic small earthquakes in a broad range of areas from Mt. Amagisan to the waters off the east coast of Izu.

Subsequently, the Izu-Oshima-kinkai Earthquake of 1978 (M 7.0) occurred in January 1978 with its hypocenter on the sea floor between Higashi-Izu Town and the Izu Oshima Island. This earthquake occurred around the southern edge of the distribution zone of the Izu Tobu Volcano Group to the northeast of the Irozaki Fault, the source of the Izu-hanto-oki Earthquake of 1974, and it can be considered to have had a larger impact on magma. However, detailed specifics have yet to be clarified.

However, what happened as a result of the Izu-Oshima-kinkai Earthquake of 1978 was critical to citizens in Ito. In November 1978, earthquake swarms began to occur off Cape Kawanazaki. This was the beginning of what later came to be called "earthquake swarms off the east coast of Izu Peninsula." There have occurred a total of 49 earthquake swarms, including small-scale ones, off Cape Kawanazaki for 36 years from the first earthquake swarm in 1978 to the end of 2014. In addition, it emerged that the center of the abnormal uplift zone moved from around Hiekawa to the southern part of the city of Ito in line with the beginning of this earthquake

The living earth of Izu (magmatism)

swarm. This location of the uplift zone was subsequently almost fixed and remains unchanged today. This resulted from magma having begun to seek a path to the surface of the earth on a full scale.

The Irozaki Fault that caused the Izu-hanto-oki Earthquake of 1974. An accumulation of fault dislocations caused the two ridges (white dashed line) to be dislocated in the same direction by the fault (white dotted line).

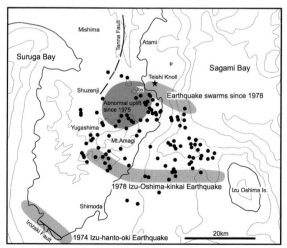

The source regions (highlighted in light gray) of primary earthquakes from 1974 to around 1978 and the abnormal uplift zone (uplift of more than ten centimeters highlighted in dark gray). The black round marks show the Izu Tobu Volcano Group. The thin solid lines show contour lines and isobaths with an interval of 400 meters.

74. The Off-Ito Submarine Eruption of 1989 (1): The process to eruption

In the previous sections, I declared that earthquake swarms and abnormal uplifts around Ito were caused by magmatism. Now that people experienced the Off-Ito Submarine Eruption of 1989, no one is doubtful about this, but of course, it had not initially been clarified that the cause of the earthquake swarms and abnormal uplifts was underground magmatism. Former Professor Hisashi Kuno at the University of Tokyo (refer to Sections 31 and 62) already argued in a paper published in 1954 that the Ito Earthquake Swarm of 1930 had been caused by magmatism. However, in the late 1970s when abnormal uplifts began to be observed around the Hiekawa area in the city of Izu, some thought that because underground faults did not cause an earthquake but slowly crept, an abnormal uplift arose. However, after earthquake swarms off the east coast of Izu Peninsula had occurred off Cape Kawanazaki almost every year since 1978 and abnormal uplifts in sync with those earthquakes had begun to be observed in the southern part of the city of Ito, it was no longer possible to simply ascribe those phenomena to faults. This is because the centers of abnormal uplifts were not the occurrence locations of earthquake swarms but always to the southwest of those places.

These observational data could be clearly explained by the theory that platy magma (dike) intruded into the underground off Cape Kawanazaki. This theory had come to be widely supported by many researchers by around 1988. Well before the Off-Ito Submarine Eruption on July 13, 1989, it had already been discovered that the cause of earthquake swarms off the coast of Ito was magmatism.

Earthquake swarms that began on June 30, 1989, were particularly massive and their hypocenter was located off the coast of the Ito Spa. At that time, I had just become an assistant professor at the Faculty of Science of Shizuoka University and one of my first students was studying geology about the Usami area in the city of Ito as the theme for graduation research. At that time, I had not yet got deeply involved in volcanological studies and just had knowledge about the Izu Tobu Volcano Group based on papers written by other researchers. In spite of this, I still well remember saying as follows to a student who was planning to go to a survey site in July: "A worst-case scenario shows that there is likely to be a submarine eruption off the coast of Usami, so be fully careful." That is, as of early 1989, the recognition that there was a great likelihood that a submarine eruption would occur off the coast of

The living earth of Izu (magmatism)

Ito was quite common at least among researchers who were interested in earthquakes and volcanoes in Izu. Now, I strongly wish this consciousness on the part of researchers had been shared in advance more broadly by local administrators and residents at that time.

GPS antennas set by the Geospatial Information Authority of Japan on top of Mt. Komuroyama. The antennas can catch minute crustal movement on the Izu Peninsula.

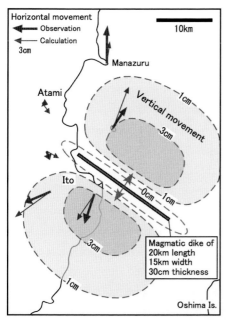

The illustration explains earthquake swarms and abnormal uplifts around Ito in 1988, the year just before the submarine eruption, by platy magma (dike) that intruded into the underground off the coast of Ito. (This was announced by two Geospatial Information Authority of Japan researchers, Tada and Hashimoto, in the spring of 1989 prior to the submarine eruption.) This theory can explain both earthquake swarms off Cape Kawanazaki and abnormal uplifts in the southern part of the city of Ito.

75. The Off-Ito Submarine Eruption of 1989 (2): Volcanic tremor and eruption

The hypocenter of massive earthquake swarm that began on June 30, 1989, was situated off the coast of the Ito Spa, the most northwestern side for earthquake swarms off the east coast of Izu Peninsula. For the depth of the seismic focus, most of them were shallow within five kilometers. On July 9, the largest earthquake with a magnitude of 5.5 occurred and caused damage to the city of Ito, including a collapse of stone fences. From the night of July 11, two days after the massive earthquake, when the earthquake swarms were almost ceasing, to the following morning, the large shakes recoded on and off by seismometers in the city of Ito greatly shocked experts who were carefully watching the situation. The recorded wave forms were not normal earthquake wave forms but the ones of volcanic tremors. A volcanic tremor is a vibration with distinctive characteristics that is considered to occur when fluids, such as magma, move underground of volcanoes. This tremor causes slower shakes than usual earthquakes and its beginning and end are unclear. The tremor often occurs immediately before and during an eruption.

The news of this tremor quickly heightened tensions and an emergency top-level meeting of the Coordinating Committee for Prediction of Volcanic Eruption was held on July 12. Until this moment, earthquake swarms off the east coast of Izu Peninsula were primarily observed and studied by seismologists and their activity evaluation was administered by the Coordinating Committee for Earthquake Prediction. That is, the meeting of the Coordinating Committee for Prediction of Volcanic Eruption whose main members were volcanologists meant that an emergency situation had arisen. After the emergency meeting, the chairperson of the Coordinating Committee for Prediction of Volcanic Eruption commented, "The tremors could be caused by underground magmatism." However, speculation spread among experts that a submarine eruption might have already begun.

On July 13, the Maritime Safety Agency (the current Japan Coast Guard) sent the survey ship *Takuyo* to the waters off the coast of Ito and launched its submarine survey of the hypocentral region of the earthquake swarms. Around 6:33 p.m., when it was dark, the crew aboard the *Takuyo* was struck by shocks and deafening sounds as if they were being hit by hammer on the bottom of the ship many times. The scared crew caught sight of a black eruption column blowing up from the surface of the sea one kilometer away. This was the very moment when the first eruption of the

The living earth of Izu (magmatism)

Izu Tobu Volcano Group had been witnessed since historic times. The eruption column was a unique one called cock's tail jet that arises when there is a phreatomagmatic explosion triggered by an encounter between water and magma. This eruption column blew up five times and the largest explosion was 230 meters in diameter and 113 meters in height from the surface of the sea. The eruption point was the spot above which the *Takuyo* had passed just seven minutes before. That is, the crew had a narrow escape from a direct hit of eruption by a hair's breadth of seven minutes. This eruption column could be seen from many places along the coast in the city of Ito and threw people who witnessed it into confusion.

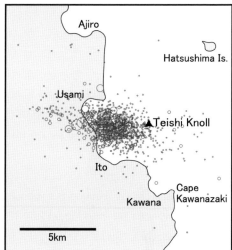

The distribution of the hypocenters of earthquake swarms from the end of June to July 1989 and the location of Teishi Knoll (the volcano that caused the Off-Ito Submarine Eruption of 1989). The data on the hypocenters are based on the earthquake catalogue of the Japan Meteorological Agency.

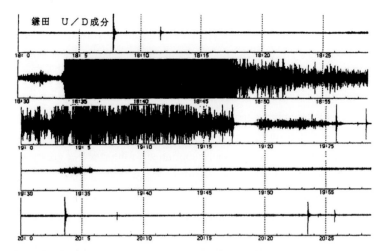

Records of volcanic tremors caused by the Off-Ito Submarine Eruption on July 13, 1989 (compiled by the Japan Meteorological Agency). The seismic amplitude of some tremors was so strong that it swung past the maximum.

76. The Off-Ito Submarine Eruption (3):
    The shock of eruption

Fortunately, the eruption column caused by the submarine eruption off the coast of Ito on July 13, 1989, was seen last around 6:44 p.m. and was not witnessed after that. That is, the eruption from the surface of the sea was seen for just ten minutes after it had begun. Massive volcanic tremors with the same scale as the ones recorded from July 11 to the following day were observed again along with the eruption on July 13, but the tremors ceased past 7:00 p.m. on that day. There were also subsequent occasional tremors, but the last time those tremors were observed was July 21.

On the sea floor 95 meters deep on which nothing was observed in a bathymetrical survey on July 9, a knoll 25 meters high had been formed on July 13. This was confirmed by the survey ship Takuyo of the Maritime Safety Agency (the current Japan Coast Guard), which had passed just above the point immediately before the eruption. After the eruption, it was confirmed that there was a crater with a diameter of 200 meters and a depth of 40 meters on the peak of this knoll. This is Teishi Knoll Volcano, which clearly remains as a submarine landform even today.

Immediately after the eruption on July 13, a large amount of black and white mix-colored pumice was found on the beach in the city of Ito. As a result of chemical analysis, it became clear that the whie part of it was created by the reheating of volcanic ash and the like contained in old strata on the sea floor. However, the black part surrounding the white part had the same chemical features as rocks ejected from other volcanoes of the Izu Tobu Volcano Group. That is, it was confirmed that Teishi Knoll Volcano was a member of the Izu Tobu Volcano Group from a petrologic perspective as well.

The Off-Ito Submarine Eruption of 1989 had a tremendous impact on not only experts but also on the local community. A special team of the local newspaper Shizuoka Shimbun energetically continued to interview a greaty variety of people who had got involved in the eruption and published long-term serialized reports about the results of their investigation from October 1989 to June 1990. Those feature articles were subsequently compiled in the form of a book entitled *Signal from the Earth* (published by the Shizuoka Shimbun) and can still be read at library and the like. It is, so to speak, a collection of dramatic stories about people who struggled hard to protect their living and safety, buffeted by the sudden awakening of

The living earth of Izu (magmatism)

a volcano.

This volcanic eruption also hugely impacted my future academic career. Around 1989, I was shifting focus from geology-centered to volcanology-centered for my research theme. I direrctly experienced the eruptions of Miyakejima Volcano in 1983 and Izu Oshima Volcano in 1986 on the ground and got strong impressions. In addition, I experienced the eruption of a volcano belonging to the Izu Tobu Volcano Group. This experience delivered the decisive blow to me and got me deeply involved in the study of eruptive history and disaster damage control.

An eruption column ejected from the Off-Ito Submarine Eruption on July 13, 1989. The picture was taken by the Maritime Safety Agency (the current Japan Coast Guard).

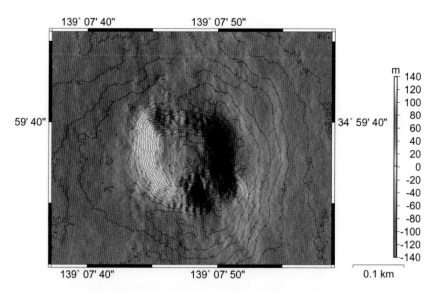

A three-dimensional submarine topographical map around Teishi Knoll on the basis of surveys after eruption (drawn up by the Maritime Safety Agency, the current Japan Coast Guard). A large crater can clearly be seen.

77. The Off-Ito Submarine Eruption of 1989 (4): Given time for preparation

Although the Off-Ito Submarine Eruption on July 13, 1989, caused a great shock and fear to the local community, the situation after the end of the eruption was a fortunate one in every way. First of all, the eruption column was seen blowing from the surface of the sea for just ten minutes or so on the day of eruption. Volcanic tremors, which can be considered to have indicated submarine eruptions, ceased in almost ten days from July 11 to 21. A rough calculation based on crustal deformation shows that magma with a volume of 20 million cubic meters rose, but just one-hundredth of it was actually ejected. In addition, magma rose on the seafloor and did not cause any damage to terrestrial areas.

As I explained in Section 58, the Izu Tobu Volcano Group has the andesitic and rhyolitic magma zone in its center and eruptions there are more likely to be massive and explosive. However, because the Off-Ito Submarine Eruption of 1989 occurred in the basaltic zone on the periphery of the volcano group, it was fortunate that the eruption was small-scale and relatively quiet. In addition, although it occurred in the basaltic zone, the submarine eruption might be explosive due to the direct encounter between a large amount of water and magma and also might cause a pyroclastic surge, or a blast involving volcanic ash, and a tsunami at the worst. In fact, however, such a serious situation did not materialize.

In addition, the Off-Ito Submarine Eruption of 1989 might linger on. For example, Surtsey Volcano is famous for its submarine eruption that is quite similar to Off-Ito Submarine Eruption of 1989. This eruption began off the coast of Iceland in November 1963. The eruption suddenly began from a submarine eruptive fissure. It initially repeated phreatomagmatic explosions and volcanic products piled up as early as on the following night, which resulted in forming an island. Subsequently, the volcano continued to erupt and created three volcanic islands and one submarine volcano in three years. On the Surtsey Island, the largest of the three volcanic islands, eruptions continued to effuse lavas onto land until 1967 and still remains as a substantial island 1,500 meters in diameter and 154 meters above sea level at its highest point.

What I explained above means that the Off-Ito Submarine Eruption of 1989 could also continue for a few years and form a new volcanic island in terms of a bad-case scenario. There was much possibility that pyroclastic surges and tsunamis would

The living earth of Izu (magmatism)

strike and inflict damage on urban areas. In addition, if the eruption had occurred not off the coast of Ito but somewhere in the terrestrial area within the Izu Tobu Volcano Group zone, there would have been some possibility that a new volcano would be formed all of a sudden in the middle of an urban area. This means that residents in Izu encountered the first volcanic eruption in approximately 2,700 years, but they were tremendously fortunate. They should wholeheartedly appreciate this luck and should also be modest enough to think that they are given time to prepare themselves for the next eruption.

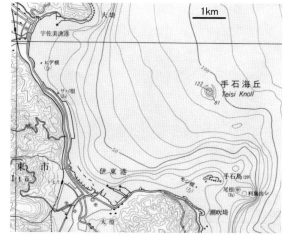

A submarine topographical map around Teishi Knoll (Maritime Safety Agency, the current Japan Coast Guard)

A bird's-eye view of the Usami area in the city of Ito and its neighboring area seen from the western side of the city. The rough location of Teishi Knoll is also shown.

Chapter 7

Coexisting with the earth

78. Predicting magmatism (1): Success in predicting its start

After the end of submarine eruptions in 1989, earthquake swarms off the east coast of Izu Peninsula still continues intermittently. For 36 years from their first earthquake swarms in November 1978 to the end of 2014, there have been a total of 49 earthquake swarms, including small ones. The Off-Ito Submarine Eruption of 1989 was part of the 20th earthquake swarm of these.

Although earthquake swarms were quiet for three and a half years after the end of the Off-Ito Submarine Eruption of 1989, there were frequent occurrences of massive earthquake swarms, including the 24th one (from May to June 1993), the 30th one (from September to October 1995), the 34th one (March 1997) and the 37th one (from April to May 1998). After these earthquake swarms, quietness seemed to have come again, but they began to occur again in May 2002 and the 44th one (from April to May 2006) was remarkably strong. Subsequently, things became quiet again, but the 46th one (from December 2009 to January 2010) occurred nearly three years after the last earthquake swarm. As these precedents show, earthquake swarms off the east coast of Izu Peninsula were considerably irregular in scale and interval and it was difficult to predict the next time of occurrence and scale.

However, as a result of steady continued efforts to observe these earthquake swarms, it has now become possible to predict some things. Particuarly significant observational data for prediction are a change of crustal strain that works on underground rocks. A strain is a numerical value that indicates a degree to which an object is deformed and reflects the intensity of stress that worked on it. A device called strainmeter to measure a strain on rock is buried underground around the Naramoto area in Higashi-Izu Town and its numerical values are constantly monitored by the Japan Meteorological Agency. As I mentioned earlier, earthquake swarms off the east coast of Izu Peninsula were caused by magma's intrusion into the underground. When magma intrudes into the underground, stress works on the crust around it and changes of crustal strain by earthquake swarms have actually been observed.

In addition, as a result of more meticulous examinations, it was discovered that a change of crustal strain began a few hours to dozens of hours prior to earthquake swarms. This is because even if crustal strains begin to change along with magma's intrusion into the underground, it does not destroy rocks at once and rocks finally begin to break after a little while and then earthquake swarms occur. By using this

Coexisting with the earth

principle, it is now possible to predict the start of earthquake swarms off the east coast of Izu Peninsula from a change of crustal strain immediately before it. As a matter of fact, for earthquake swarm that began on the evening of December 17, 2009, when the earthquake swarm would start had already been predicted as of midnight on December 16.

Swarm number	Date of beginning	Duration (days)	Number of felt earthquakes	Magnitude of the largest earthquake	Maximum JMA intensity
1	November 23, 1978	73	26	5.5	4
4	June 23, 1980	101	235	6.7	5
8	January 14, 1983	23	47	4.6	3
9	August 30, 1984	43	95	4.7	3
11	October 13, 1985	31	12	4.1	3
13	October 10, 1986	23	16	4.8	3
14	May 6, 1987	33	90	5.1	3
15	February 14, 1988	18	8	4.7	3
18	July 26, 1988	52	289	5.2	4
20	June 30, 1989	69	494	5.5	4
23	January 10, 1993	9	38	4.2	3
24	May 26, 1993	21	174	4.8	4
30	September 29, 1995	30	153	5.0	4
32	October 15, 1996	27	43	4.3	4
34	March 3, 1997	24	449	5.9	5 lower
37	April 20, 1998	44	211	5.9	4
44	April 17, 2006	26	49	5.8	5 lower
46	December 17, 2009	27	257	5.1	5 lower

Main earthquake swarms off the east coast of Izu Peninsula that occurred in and after November 1978. Earthquake swarms whose largest seismic intensity was more than 3 on the Japanese seismic scale are listed. The chart is based on data from the Japan Meteorological Agency.

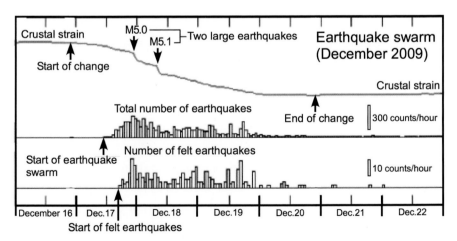

Earthquake swarm occurred in December 2009 and associated change in crustal strain. The chart is based on data from the Japan Meteorological Agency.

79. Predicting magmatism (2): Predicting scale and end

It has recently become clear that observational data concerning crustal strains on underground rock can be utilized not only to predict the beginning of earthquake swarms off the east coast of Izu Peninsula but also to predict their scale and days of continuation and to examine the possibility of voncanic eruption. A notable point in this case is the amount of change of crustal strain. This numerical value is considered to reflect the amount of magma that intruded into a shallow space underground. The larger the amount of magma that intrudes into the underground is, the stronger the stress that works on the surrounding rocks is and the more the crustal strain increases. Consequently, the number of earthquakes induced by crustal strains rises and days of continuation get longer and it can be considered that the possibility of magma reaching the surface of the ground will also grow.

In fact, the total change of crustal strain triggered by earthquake swarms (the 20th one) that caused submarine eruptions in 1989 was four times as large as that of earthquake swarm (the 46th one) that continued from December 2009 to January 2010. Other earthquake swarms that marked similar huge changes of crustal strain include the 18th one (from July to September 1988), the 30th one (from September to October 1995) and the 37th one (from April to June 1998). The days of continuation recorded with these four earthquake swarms were long, 30 to 69 days, and the number of earthquakes (including unfelt earthquakes measured by the seismometer set at the Kamada area in the city of Ito) was also massive, 9,000 to 25,000 times. In contrast, in the case of earthquake swarm that occurred from December 2009 to January 2010, they were obviously small-scale with the days of continuation being 27 days and the number of earthquakes being 6,525 times.

However, it is necessary to note that it is after earthquake swarm have almost come to an end that the total change of crustal strain is determined. This makes prediction too late and useless. Becaese of this, although lower in the level of precision, the method is adopted of making predictions on the basis of the numerical value measured when the rate of change of crustal strain (the momentum with which magma intrudes into the underground) during earthquake swarms hits the maximum. This method enables you to broadly estimate the scale and the time of ending within a few days from the beginning of earthquake swarms. For the change of crustal strain triggered by earthquake swarm that occurred from December 2009 to January 2010, it was possible to make predictions based on the rate of change of crustal

Coexisting with the earth

strain as of December 20, three days after the start of the earthquake swarms. The prediction results were ten days or so for the days of continuation and approximately 2,600 times for the number of earthquakes. In actuality, however, the days of continuation were 27 days and the number of earthquakes was 6,525 times, as mentioned above. Both were larger than the predicted values, but in the case of earthquake swarm that continued from December 2009 to January 2010, the prediction gaps seemed to have been affected by an overall shallow hypocenter. After all, you should evaluate the fact that it became possible to roughly estimate the scale and duration of earthquake swarms at the early stage.

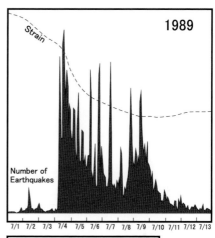
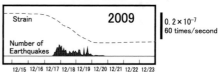

A comparison of the number of earthquakes (per hour) and changes of crustal strain on the same scale with a focus on earthquake swarm in 1989 (above), when the Off-Ito Submarine Eruption of 1989 occurred, and earthquake swarm in 2009 (below). The horizontal axis means dates. The illustration is based on the observations by the Japan Meteorological Agency.

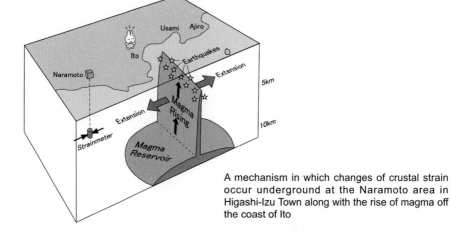

A mechanism in which changes of crustal strain occur underground at the Naramoto area in Higashi-Izu Town along with the rise of magma off the coast of Ito

80. Predicting magmatism (3):
Volcano monitoring and disaster damage control system

Observational results since 1978 show that there occurred several earthquake swarms off the east coast of Izu Peninsula involving massive strain changes and that one of those earthquake swarms (from June to September 1989) caused submarine eruptions. It was the one and only case that led to an eruption compared with all earthquake swarms to this day. For this case, the total number of earthquakes (including unfelt earthquakes) amounted to almost 25,000 times and the number of felt earthquakes was 494 times. Both values marked the largest on record. That is, it seems that if you focus on the level of strain change and the number of earthquakes, you can assess the possibility of eruptions. In the meantime, public systems are necessary to actually utilize such prediction technology for disaster damage control.

To develop this system and officialize it for the local society, the Committee on Disaster Mitigation Measures for Izu Tobu Volcano Group was established in January 2009 by the Shizuoka prefecture. This Committee was jointed by the Ito and Izu municipal governments, as well as disaster damage control-related organizations, such as the Japan Meteorological Agency. Its final report was announced in March 2011. Based on this report, the Japan Meteorological Agency began to apply information on estimated seismic activity and the Volcanic Alert Levels (refer to Section 81) to the Izu Tobu Volcano Group on March 31, 2011. In addition, the Shizuoka prefectural government and the Ito and Izu municipal governments revised their respective local disaster damage control plans.

The final report of the aforementioned Committee includes a map illustrating the "zone of possible eruption" and the "zone of possible eruption damages." This map is based on the assumption that the current occurrence area of earthquake swarms will remain unchanged in the future. However, it can be said that the prototype of the Izu Tobu Volcano Group Hazard Map has finally appeared, although I deplored in the past that such system was not put in place. This identification of particular hazardous areas in case of potential eruption made it possible to draw up specific disaster damage control plans, including evacuation procedures. The above-mentioned Committee temporarily disbanded, but in March 2012, the Volcano Disaster Management Council for Izu Tobu Volcano Group was reorganized with the participation of almost the same members and a leading role from the Ito municipal government. Currently, this Council is deliberating specific evacuation plans.

Coexisting with the earth

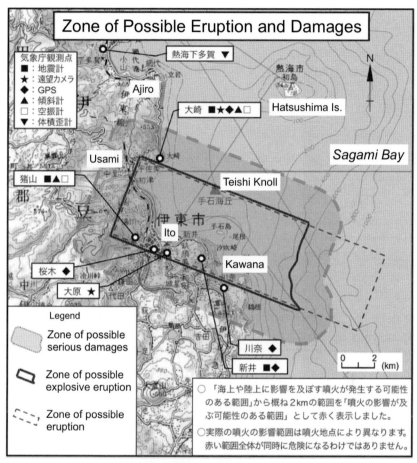

A map illustrating the "zone of possible eruption" and the "zone of possible eruption damages" in relation to the Izu Tobu Volcano Group (drawn up by the Japan Meteorological Agency and partly translated by the author)

81. Predicting magmatism (4): Magmatic activity scenario

A new disaster damage control system introduced for the Izu Tobu Volcano Group is also epoch-making in that it is based on the "Magmatic Activity Scenario," which was drawn up according to the specific cases of preceding numerous earthquake swarms and eruptions that occurred within the hazardous zone. In addition, the Japan Meteorological Agency has introduced two information delivery systems related to disaster damage control on the basis of the magmatic activity scenario: (1) information on estimated seismic activity; and (2) the Volcanic Alert and the Volcanic Alert Levels.

The illustration of magmatic activity scenario is set on a time span from the left to the right. The "Normal" on the far left shows usual conditions. It can be said that there is almost no possibility of eruption under the conditions. Unlike many other volcanoes, the Izu Tobu Volcano Group has no fixed craters because it has no stable pathways to the surface of the earth. An eruption in 1989 created a new crater (Teishi Knoll) and the pathway to the crater was temporarily formed. However, it is conceivable that because more than 20 years passed, the pathway cooled and closed.

At the initial stage of earthquake swarms, the hypocenter moves from a deep spot to a shallow spot ("Magma starts to rise" in the illustration of the magmatic activity scenario). This is because magma rises, breaking underground bedrocks and causing earthquakes. The magma that rose stays at a somewhat deep spot nine to six kilometers underground ("Magma stopped at >6km in depth" in the illustration of the magmatic activity scenario) or goes up to a shallow spot six to three kilometers underground ("Magma reaches a depth of 3-6km" in the illustration of the magmatic activity scenario). Although the former also causes earthquake swarms, the latter can cause more active ones. The latter, close to the surface of the earth, can cause damage through shakes. In this case, the Japan Meteorological Agency announces information on estimated seismic activity. Historical experiences since 1978 tell that the probability of both cases is almost fifty-fifty.

For example, information on estimated seismic activity is announced in the following way: "Specific prediction of earthquake activity on the basis of observational data as of XX (a specific date and time) is as follows: The seismic scale and intensity: a magnitude of 5; and 4 to 5-lower on the Japanese seismic scale; the number of earthquakes registering more than 1 on the Japanese seismic scale: 200 to 400 times; and duration of activity: several days."

In addition, when some magma that rose to a shallow spot sometimes goes up to a further shallower spot two to one kilometers underground ("Magma reaches a depth of <3km" in the illustration of the magmatic activity scenario). The probability of this scenario is as low as 5%. However, low-frequency earthquakes, or volcanic tremors are observed and the possibility of eruptions sharply rises. Therefore, the Volcanic Alert Levels are raised and the Volcanic Alert is announced. In addition, municipal governments issue an evacuation advisory or order.

The possibility of eruptions is also fifty-fifty at this stage. More specifically, seen from the starting point of earthquake swarms, the possibility of eruptions is 2 to 3%, which can be obtained by multiplying the 5% mentioned above by 50%. If an eruption finally occurs, eruptive phenomena with different characteristcis arise according to the difference in eruption place (land or sea) ("Submarine eruption" and "Subaerial eruption" in the illustration of the magmatic activity scenario).

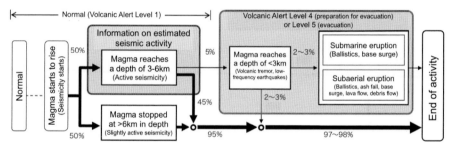

The magmatic activity scenario for the Izu Tobu Volcano Group (drawn up by the Japan Meteorological Agency and translated by the author). The illustration shows the possibility of each simulated situation seen from the starting point of earthquake swarms.

82. Learning volcanoes

The magmatism of the Izu Tobu Volcano Group has been active off the coast of Ito since November 1978 and for several years prior to that, earthquake swarms occurred in many parts of the Izu Peninsula, including Mt. Amagisan. (refer to Section 73). Basically, the Izu Tobu Volcano Group is distributed in a broad range of areas encompassing the city of Ito, the city of Izu, the city of Izunokuni, Higashi-Izu Town and Kawazu Town. That is, it is not unnatural that magmatism will affect any of these five cities and towns at any time. In addition, areas other than these cities and towns may also be damaged by volcanic ash, debris flows and tsunamis triggered by eruptions. That is, disaster damage control measures for the Izu Tobu Volcano Group are an issue that should be tackled by the whole of Izu.

The fist step toward realizing this basic ideal is that the entire residents of Izu need to recognize the necessity of acquiring knowledge about volcanoes on a daily basis and developing a volcano-conscious lifestyle. What is siginificant in practicing this approach is that the citizes in Izu have the recognition that volcanic knowledge and information do not have a negative effect on their living and tourism but also have a positive effect if they are put to their proper use. In fact, when an earthquake swarm begin, many tourists often cancel visiting places located in or near the earthquake zone even if they are not very serious. This is because those people are short of knowledge about volcanoes. Ignorance causes them to be excessively afraid of volcanoes.

However, I would like the citizens in Izu to think about how much they know about volcanoes before criticizing those excessive reactions from tourists. It is natural that tourists be greatly anxious about or afraid of what local residents are unfamiliar with. If local citizens make an active effort to obtain knowledge about volcanoes and break away from their baseless anxiety and mistaken images, they can learn to be properly afraid of volcanoes and properly cope with emergencies. Safe and reassuring tourist spots where substantial volcanic measures are secured by keenly aware local citizens and administrators are likely to attract many sightseers just because of that. In addition, it is essential to realize that volcanoes that created the earth of Izu provide many tremendous benefits for local residents and tourists. I will explain this in detail in the next section and beyond.

Coexisting with the earth

An open seminar about volcanoes held in the city of Ito in November 2009. I gave a lecture at this session. These semilars to learn volcanoes are often held. The picture was provided by the Ito Board of Eduction.

I gave an outdoor volcanic lecture for Izu Sogo High School students in June 2010.

83. The blessings of volcano (1):
Volcanoes create land

Volcanoes can create flat land. This can be said to be the most significant blessing of volcanoes, but it is often forgotten. For example, suppose that the Izu Tobu Volcano Group did not exist, what would the landform of the area be like? This simulation highlights how great the blessings of the Izu Tobu Volcano Group are. As I explained in Section 50, the lavas effused from Omuroyama Volcano filled in the bumpy landform around it and created the Izukogen Plateau. The lavas also poured into the Sagami Bay and created the Jogasaki Coast (see Frontispiece 1). A bird's-eye view of the Jogasaki Coast clearly shows that the edge of lava flows divided into branches toward the sea like a sago palm leaf and expanded terrestrial areas. Please imagine how the landform would look if the lava flows were removed from the picture. The Jogasaki Coast and the flat land behind it would disappear and the coastline would recede by more than two kilometers. In addition, the Izukogen Plateau would turn into a steep mountainous area. If that had happened, it would have been impossible to build the current villa resorts that get plenty of sunshine.

In addition to Omuroyama Volcano, many volcanoes, such as Komuroyama Volcano, Umenokidaira Volcano, Kadono Volcano and Ogi Volcano, effused lavas one after another and conducted major land creation works to make the flatland of southern Ito. Without these land creation works, the city of Ito would have become an isolated village surrounded by the steep slope of mountains and the coast and would not have developed as it does today. Thanks to the eruptions of the Izu Tobu Volcano Group and the lavas effused from those eruptions, a spacious area with plenty of sunshine was created and became a place where many citizens and tourists gather.

Many other older volcanoes, as well as the Izu Tobu Volcano Group, unceasingly continued these activities beneficial to local residents and societies. The spacious valleys and plains through which the Kanogawa River and the Omigawa River run were continuously provided with a large amount of rock debris by large terrestrial volcanoes (refer to Sections 29 and 30), including Amagi Volcano, as well as the Izu Tobu Volcano Group. Dairy farmers work in highlands around the boundary between the city of Izu and Nishi-Izu Town by utilizing their flat landform, which was created by the lava flows spouted from Nekko Volcano (refer to Section 34). A spacious hilly area in the western part of the city of Izu, where the Shuzenji Natural Park, Shuzenji

Coexisting with the earth

Nijino Sato Park and golf courses, such as Laforet Shuzenji, are located, was created by the lava flows effused from Daruma Volcano. A small plateau, in which day lilies are blossoming, to the west of Cape Irozaki in southernmost Izu, was also created by the lava flow effused from Nanzaki Volcano (refer to Section 35). There is a list of uncountable similar cases.

A bird's-eye view of the Izukogen Plateau and Omuroyama Volcano. The small pudding-shaped hill at the top of the picture is Omuroyama Volcano. The plateau stretching in front of this volcano is the Izukogen Plateau. The lava flows effused from Omuroyama Volcano ran into the rugged coast (Futo Coast) at the front.

84. The blessings of volcano (2):
Volcanoes create water sources

Volcanoes also give the blessings of water. Lake Ippekiko, situated on the southern side of the Ito Spa, was created as a result of water having accumulated in the crater of eruption approximately 100,000 years ago (refer to Section 40) and provided us with a perfect relaxation place (see Frontispiece 2 [below]). A depression to the southeast of the lake is also a crater (Numaike) formed by the same fissure eruption as Lake Ippekiko. Currently, this area is a housing lot and a marsh, but it used to be a lake.

The lava flows effused from Omuroyama Volcano 4,000 years ago to the south blocked the exit of a ravine on the eastern side of the Rokurobatoge Pass and created a lake twice as large as Lake Ippekiko (refer to Section 50). The remnant of this lake is the current Ike Basin in the city of Ito and good-quality rice is cultivated in the area by utilizing the features of the marsh.

Volcanic products often involve many fractures and openings. Underground water sometimes accumulates in there and springs up halfway and at the foot of the volcano. A huge amount of underground water that accumulated in the fractures of the lava flows effused from Amagi Volcano (refer to Section 32) spouted and ran through many mountain streams. There are many wasabi (Japanese horse radish) fields in those areas. Rich water running through the terraced paddy fields in the Ishibu area, Matsuzaki Town, also sprang out from the fractures of lavas effused from upstream Jaishi Volcano (refer to Section 35).

Such underground water soaks deep into the underground together with seawater and is heated by high-temperature geothermal heat of the earth of Izu. The underground water gets various ingredients from rocks and then gushes out to the ground again. It is a hot spring, which is a significant feature of Izu tourism. This high-temperature geothermal heat was carried by magma that created many volcanoes in Izu and is, so to speak, the "remaining heat" of volcanic activity.

High-quality spring water peculiar to volcano is utilized for the source of water supply in the city of Ito, for example, and in the case of the Chaine des Puys Volcano Group in France, its water is exported as the mineral water "Volvic" to all around the world. I hope that the readers will know that not only hot springs but also spring water in Izu has great potential brand value.

Coexisting with the earth

A view of terraced paddy fields seen in the Ishibu area, Matsuzaki Town. The picture was taken by Yusuke Suzuki.

A hot spring well that is often seen on the Izu Peninsula

85. The blessings of volcano (3):
Volcanoes produce stone and tourist resources

Stone and mineral resources produced by volcanoes have been mined since a long time ago and have been used for our living. For example, various ingredients precipitate from magma and hot springs underground volcanoes and become mineral deposits. Gold used to be mined in many parts of Izu. The Izu Silica Stone in the Ugusu area, Nishi-Izu Town (refer to Section 36), once occupied a large share of materials for plate glass in our country for a long while. Obsidian produced around the Kashiwatoge Pass on the boundary between the city of Ito and the city of Izu is also one of the glass resources created by volcanoes and was utilized as material for arrowhead and knife in the Jomon period (refer to Section 37).

Andesitic lavas effused from terrestrial large volcanoes (refer to Sections 29 and 30) in Izu, such as Taga Volcano, Usami Volcano, Amagi Volcano and Daruma Volcano, were cut in a large amount as material stone to build Edo Castle and were transported by ship. There remain many stone pits (called "ishichoba") in mountains and on coasts. Some rocks discovered at those old remains have arrow holes (traces of processing) and engraved marks of daimyos (feudal lords) who supervised rock-cutting. In addition to these lavas, tuff created by submarine volcanoes when most of Izu was submerged in the sea was also mined as stone material a long time ago. Tuff is formed as a result of volcanic ash and pumice hardening after a long while. Scoria (dark-colored pumice) ejected from the Izu Tobu Volcano Group was also widely utilized as aggregate to be mixed with concrete and was mined in many areas.

Volcanoes also create beautiful objects. Tuff mentioned above sometimes has beautiful patterns formed by sea currents and waves. The cliff on the Dogashima Coast in Nishi-Izu Town is an example of it (refer to Sections 14 and 15). The beautiful shape of Omuroyama Volcano is called scoria cone peculiar to volcano (refer to Section 48) (see Frontispiece 2 [above]) and the lavas effused from the volcano created uniquely shaped objects, such as pothole and scoria raft (refer to Section 51) (see Frontispiece 4 [below]). Several lava flows effused from the Izu Tobu Volcano Group poured into the mountain streams associated with Amagi Volcano and created spectacular falls, such as the Jorennotaki Falls (the city of Izu) and the Kawazu Nanadaru Falls (Kawazu Town) (see Frontispiece 7 [above]). Some of those lava flows have cooled joints of beautiful arrangement, such as columnar joints and platy joints (refer to Sections 45 and 46) (see Frontispiece 4 [above]).

Coexisting with the earth

1. Broad and flat land
Lava and debris flows bury bumpy landform and create plateaus and plains, which are useful to various human activities.
Examples: Jogasaki Coast, Numazu Plain, and Mishima Fan
2. Beautiful mountains and plateaus
Volcanoes make beautiful mountains with gentle slopes.
Examples: Mt.Omuroyama, Mt.Fuji, and Amagikogen Plateau
3. Lakes
Lava flows often dam valleys and create lakes. Eruptions sometimes make crater lakes.
Examples: Ike Basin and Lake Ippekiko
4. Groundwater
Volcanic products often involve many fissures and openings. Underground water accumulates in there and springs up at the edges of them.
Examples: Mt.Suidoyama of Ito, and springs at Mishima and Mt.Amagisan
5. Characteristic geomorphology and moldings
Volcanic ash sometimes makes beautiful strata. Lava flows create beautiful coasts and columnar joints.
Examples: Dogashima and Jogasaki Coasts
6. Fertile soil
Volcanic ash changes to fertile soil and nourishes forests and farms.
Examples: Forest of Mt.Amagisan and farms at Kannami Town
7. Ores and quarries
Various ingredients precipitate from magma and hot springs underground volcanoes and become ores. Lava flows, tuff, and scoria are useful as building materials.
Examples: Various ores and quarries in the whole Izu Peninsula
8. Geothermal energy and hot springs
Geothermal energy originated from magma makes hot springs, which are used for spa resorts and sometimes for generating electric power.
Examples: Hot springs in the whole Izu Peninsula

A list of typical blessings of volcano

Muroiwado in Matsuzaki Town. This used to be a stone pit called ishichoba from which people obtained tuff as stone material. The picture was taken by Yusuke Suzuki.

86. The dream of geopark (1): Mother earth

Many blessings volcanoes in Izu teach us that volcanoes are a motherlike natural existence for residents and visitors. Ever since old times, people in Izu have ingeniously utilized landforms, volcanic products, spring water, hot springs, mineral deposits and stone that volcanic activities provided to many parts of the region for living spaces and tools. We have naturally acquired the techniques to coexist with the earth. We cannot be easily conscious of these natural blessings in our daily lives, but if you have just a little knowledge, you can find many signs of the breathing of the earth in your familiar environments.

In the ever-lasting activity of the earth, blessings and disasters are always two sides of the same coin. In the long term, the Izu Tobu Volcano Group will continue to erupt from now onward as well. However, in the whole life of most volcanoes, the eruptive period is just a quick moment and the dormant period is even longer. An estimate based on the eruptive history introduced in Sections 56 to 58 shows that the Izu Tobu Volcano Group has just erupted at the average interval of 3,000 years over the past 30,000 years. The submarine eruption in 1989 (refer to Sections 74 to 77) was really an unfortunate event. Therefore, it is a mistake to excessively fear and dislike the Izu Tobu Volcano Group. As long as substantial preparations and measures for eruptions and disasters based on a worst-case scenario are in place, residents in Izu and tourists will continue to be able to enjoy the blessings of the earth without worry.

Now, a new approach to regional promotion highlighting those natural assets provided by the earth is being globally proposed and carried out. This approach is represented by the concept of geopark that the United Nations Educational, Scientific and Cultural Organization (UNESCO) started to support. The word *geo* means the earth and needless to say, the word park means a public park. Geopark is a new system under which a local society with a rich pool of significant natural assets provided by the earth (*geo*) develops its economic and cultural activities by preserving and utilizing those assets in an effort to eventually boost regional promotion.

The economic activities in this context primarily mean tourism and its related industries, encompassing not only attracting tourists and coordinating sightseeing tours but also the development of *geo*-related tourist products, projects for the preservation and disaster management of geopark tourist spots (geosite) and the development of local specialties and gift items associated with the concept of *geo*.

In addition, the cultural activities in this context encompass personnel development, such as tourist guides (geoguide), education for next-generation young people who support geopark and encounraging them to return to their home towns, the creation and distributon of art works reflecting the themes of geopark, study of geosites and the development of new geosites. If these activities work properly, they are expected to be effective for revitalizing and promoting local communities.

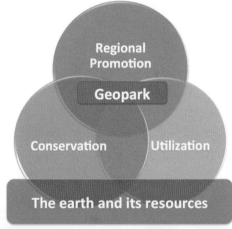

The system and purpose of geopark

Local high school students are teaching local elementary school students how the landform of Omuroyama Volcano was created and its characteristics. This is part of geopark education.

87. The dream of geopark (2): The assets of geopark

People often associate the name of geopark with strata and rocks as its component assets. However, if you simplistically regard geopark as "geologic assets" and the "geologic version of World Heritage," you get it wrong. The assets of geopark encompass everything existing on the earth.

Local communities have their own distinctive strata and rocks. These natural products provide local natural condictions, such as landform, soil, underground water and hot spring, and grow terrestrial and marine animals and plants. These conditions produced a great variety of local specialties. Mining and manufacturing industries, agricultural, forestry and fisheries industrieries and tourist industries based on those local assets have long been developed. In the meantime, the earth sometimes showed its fangs and attacked human beings, causing various natural disasters. However, seen from a long-term perspective, such diasters created significant landforms and resources and helped people develop disaster damage control science and technology and nurture knowledge and culture related to how to coexist with disasters. These natural blessings and threats resulted in creating local unique landscapes, industries, history, human resources, science, technologies, cultures, thoughts, beliefs, arts and education. These are all the assets of geopark.

In other words, at the root of everything that exists in local communities are Mother Earth and stories leading to it. Geopark is the stage on which local residents acquire and enjoy this knowledge and discovery and promote local economic and cultural activities by preserving and utilizing those resources. This is not difficult at all. For example, all of your familiar beautiful landscapes, things and local activities have meanings and stories that created them. If you find the truths behind those local things, you will see a whole new world stretching in front of you. When you discover that all things that you thought had nothing to do with each other are actually closely associated with each other, you will be profoundly moved and touched. Geopark is a system that gives all people who love a particular region, including its residents and visitors, the opportunity to enjoy such an impressive experience.

If I focus on tourism here, conventional tourism disregarded these local fundamental origins and provided just a part of them, that is to say, only beautiful scenery, food and hot springs without paying keen attention to their core meanings. In striking contrast, geopark-based tourism (geotourism) focuses on all tangible and intangible things associated with the earth and the relationships between those things. Even things that have been thought of as quite common, ordinary and unattractive until today can be significant component assets of geopark if their relationships with the history and activity of the earth are clarified. This enables a certain place and area that have been considered completely unattractive to this day to suddenly attract global attention. That is, geopark can be said to be a spectacular activity to restore regional value and pride.

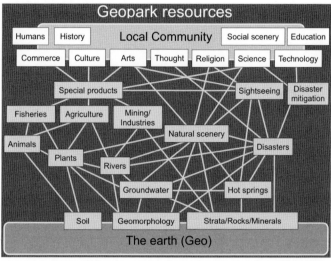

The component assets of geopark. Geopark-based tourism (geotourism) focuses on all these things.

A geopark signboard explaining how the natural landscapes of major geosites were created. An example of Cape Koganezaki in Nishi-Izu Town.

88. The dream of geopark (3):
The current situation of Izu Peninsula Geopark

Geopark is categorized into Global Geoparks and National Geoparks. Global Geoparks are designated by the UNESCO-supported Global Geoparks Network (founded in 2004). National Geoparks are classified as the lower category of Global Geoparks and, in the case of Japan, are desiganeted by the Japan Geopark Committee (founded in 2008).

As of the end of 2014, 111 regions in 32 countries are designated as Global Geopark sites. In Japan, 36 regions are designated as National Geopark sites and seven of those regions (Unzen Volcanic Area, Toya Caldera and Usu Volcano, Itoigawa, San'in Kaigan, Muroto, Oki Islands and Aso) are also designated as Global Geopark sites.

The Izu Peninsula was designated as a National Geopark site in September 2012. It is managed by the Izu Peninsula Geopark Promotion Council, which was established jointly by the Shizuoka prefectural government and seven local cities and eight local towns. Seven full-time officials, including two researchers, are dedicated to carrying out various operations. There are more than 100 signboards explaining significant tourist destinations (geosites) and promenades have also been built. In addition, geopark visitor centers, although small, have been opened in many places of the peninsula and it has also been determined that the central museum will be opened in 2016.

Furthermore, more and more local people are steadily supporting the geopark project. Plans to nurture geoguides who understand the landscapes and culture of Izu are being implemented and there are now more than 100 certified geoguides. They have been coordinating a great variety of tours, such as walking, cruising by ship and diving, and those tours are receiving favorable responses from participants. Some geoguides play an active role as local disaster damage control leaders, utilizing their specialized knowledge. Local schools are becoming more active in geopark education. In addition, local citizens create the special snack food "Geogashi" featuring strata and rocks and hold traditional flower arrangement exhibitions "Geoikebana." These cultural activities are attracting international attention and are being highly evaluated. In this way, every Izu enthusiast can participate in the geopark project by connecting their own professional and personal skills with the concept of geopark and all these activities define the value of geopark. It can be said that the geopark project is a higher-level activity than the World Heritage project that is managed by top-down decision-making.

The Izu Peninsula has finally been nominated as a candidate for Global Geoparks in recognition of these many assets and commitments. The application will be on-site examined in the summer of 2015 and its result will be announced in Setember 2015.

An example of geotour featuring a cruise by ship (Jogasaki Coast in the city of Ito)

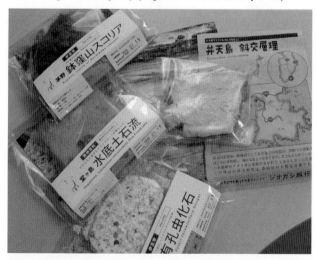

Special baked snack food "Geogashi" featuring strata and rocks of the Izu Peninsula

Geohistory of the Izu Peninsula

伊豆の大地の物語 英語版

2015年3月29日 初版発行

著　者／小山　真人　Masato Koyama
翻訳者／平井　和也　Kazuya Hirai
発行者／大石　　剛　Go Oishi
発行所／静岡新聞社　The Shizuoka Shimbun
　　　　〒422-8033　静岡市駿河区登呂3-1-1
　　　　電話　054-284-1666
印刷・製本　図書印刷　TOSHO Printing Co.,Ltd.